Lyman Belding

Land Birds of the Pacific District

Lyman Belding

Land Birds of the Pacific District

ISBN/EAN: 9783744714648

Printed in Europe, USA, Canada, Australia, Japan

Cover: Foto ©berggeist007 / pixelio.de

More available books at **www.hansebooks.com**

Occasional Papers

OF THE

CALIFORNIA
ACADEMY OF SCIENCES.

II.

SAN FRANCISCO,

1890.

LAND BIRDS

OF THE

Pacific District

BY

LYMAN BELDING.

SAN FRANCISCO:
CALIFORNIA ACADEMY OF SCIENCES,
September, 1890.

Committee of Publication:

H. W. HARKNESS, GEORGE HEWSTON.
T. S. BRANDEGEE, H. H. BEHR.

Editor: TOWNSHEND STITH BRANDEGEE.

LAND BIRDS OF THE PACIFIC DISTRICT.

In the fall of 1883 the leading ornithologists of the United States and Canada met at New York and organized the American Ornithologists' Union, appointed committees for the revision of the classification and nomenclature of North American birds, on their migration, avian anatomy, the elegibility of the European house sparrow in America, and on faunal areas.

The United States and British North America were divided into thirteen districts, including the light-houses of the coasts as one, with a superintendent for each district.

The Pacific District comprises California, Oregon, Washington and Nevada, in which I superintended the collection of data concerning migration and distribution of the birds for about two years, resigning early in 1886, as I thought we had already accomplished about all we were likely to in this line of inquiry, with the limited number of observers. The report from the District of British Columbia, Mr. John Fannin, Superintendent, was, at the suggestion of Dr. C. Hart Merriam, Chairman of the Committee on Migration, and by the courtesy of Mr. Fannin, incorporated in our report.

Mr. Fannin informed me that his list of British Columbia birds was made up partly from his own notes and partly from those published by John K. Lord, and says that having traveled the province pretty extensively, he has ascertained that Mr. Lord made some mistakes in limiting the range of some species, and in recording the habits of others, and that the observations concerning British Columbia must be taken as referring to the whole province, and not for any particular district, unless otherwise stated.

"British Columbia," he writes, "is a country of

migrants within itself, for while it is winter in one portion, the flowers of summer are blooming in another. It is also a country of great distances, and it is very wonderful what a change in our bird fauna may be seen in traveling a few miles. It is only nine miles from Burrard Inlet to New Westminister, and yet of our summer visitors, there are some four or five species which are abundant at the latter place that are entirely unknown at the former; and so it is all through the Province, although the difference may not be so marked."

The notes of the light-house keepers on the coast of British Columbia and Washington were kindly forwarded by Dr. Merriam for the same purpose.

In 1885 the Division of Economic Ornithology was established in connection with the U. S. Department of Agriculture, at Washington, with Dr. Merriam as ornithologist and Dr. A. K. Fisher assistant ornithologist. The work on migration and distribution was continued here, and observations on the food habits of the birds were added; and still later, observations on the food of mammals became a part of the work. This report was mostly type-written by that department, proof read by Dr. Fisher, and the most of the data of 1886 and later was incorporated by him. To do this, he gave very valuable time and much intelligent labor without the hope of reward, although he was, during this time, superintendent of the extensive and important Atlantic District. Finally, we are placed under still greater obligations to the Department of Agriculture by the return, at my request, of the type-written copy.

This report aims, mainly, to show the arrivals and departures of migrating species, as well as to give a catalogue of all the species known to occur in the district. The data on the food of the birds has been mostly eliminated from the report, being usually meagre and, therefore, unsatisfactory. It will, however, be preserved for future use. All persons are credited with the data which they contributed to the report, which has been simply a labor of love from beginning to end.

In California the spring migrating land birds almost invariably come gradually and almost imperceptibly, probably many arriving in the night or early morning; and I have never seen, on the Pacific Coast, what might properly be termed a "bird wave," except upon one occasion, and that was in the spring of 1866, before I began to keep a record of bird movements. Mr. F. Stephens noticed a "bird wave" at Campo, San Diego County, the last of April, 1877 (Bull. Nutt. Orn. Club, July, 1883, p. 188). Except these, I know of no other occurrences of this kind on this Coast, of course excluding the occasional movements of large bodies of geese and ducks. It is very difficult, therefore, to ascertain the lines of flight, but we may conclude that many summer residents of California and northward bear well to the east in seeking their winter quarters in the tropics, as at least a dozen species which breed in California have not been observed much south of San Diego at any time, among these being such conspicuous species as Bullock's oriole and the Arkansas flycatcher, which, in common with the other species, occur far south on the mainland in Mexico in winter. Then there are summer residents of Washington, British Columbia and Alaska which have never been collected in California, though they spend the winter in the tropics. The high Sierra Nevada and Cascade ranges would seem to be an impassible barrier to the migrants, and they undoubtedly have great influence in determining lines of flight—but not nearly as much as would naturally be supposed, as a number of species which winter in California migrate vertically and spend the summer on the east slope.

I have seen pelicans, geese, Lewis's woodpecker and others crossing between eight and nine thousand feet above sea level. Mountain quail cross on foot, some of them making a journey of sixty or eighty miles, no matter how deep the snow in spring, returning in fall, sometimes over a foot or two of snow on the divide, leaving their summer resorts where there is no snow, to reach their well known winter home. A great many of the summer visitants enter the San Joaquin Valley by the Tehachapi Pass, altitude 4,000 feet. Col. N. A. Goss, in the spring of 1884, noticed terns at Julian, San Diego County, altitude 4,500 feet, "crossing from the Gulf of California to the Pacific." Mr. Walter J. Morgan saw an immense migration of sand-hill cranes and geese which lasted about two weeks, by day and night, from Ensenada, Lower California, to Port San Felipe, on the Gulf, in Oct. and Nov., 1884, much of the Peninsula between these localities being no less than 4,000 feet above sea level.

Probably a great many small birds take nearly the same course in fall, cross near the head of the Gulf and spend the winter in Mexico. Mr. F. Stephens, who has collected a long time in San Bernardino Valley, says the spring migrants enter that valley from the southeast and return in an opposite direction in fall. He thinks a great many cross the Gulf of California from Mexico.

As the names of all correspondents are connected with the information they furnished, it is not necessary to name them elsewhere.

The authorities quoted, sparingly in most instances, but in their own language, usually, or a part of it, are:

Dr. J. S. Newberry. Explorations and Surveys for a Railroad Route from the Mississippi Valley to the Pacific Ocean. Vol. 6, part 4, number 2, War Department.

S. F. Baird, John Cassin and Geo. N. Lawrence. Explorations and Surveys, etc. Vol. 9, War Dept. (often quoted as Vol. 9, Pacific R. R. Repts.).

Dr. A. L. Heermann. Explorations, etc. Vol. 10, No. 2, War Dept. (Vol. 10, Pacific R. R. Reports.).

Drs. G. Suckley and J. G. Cooper. Explorations, etc. Vol. 12, Book 2, part 3, War Dept. (reissued as Natural History of Washington Territory), the extracts from their respective notes being prefaced in this report with Cooper, 1860; Suckley, 1860, this being the date of the publication of their work.

Dr. J. G. Cooper. Geological Survey of California, J. D. Whitney, State Geologist. Vol. 1, Land Birds, edited by S. F. Baird from manuscript and notes of Dr. J. G. Cooper. Herein cited as Cooper, 1870. [Auk, vii, 214.—W. E. B.]

II. W. Henshaw. Wheeler's Topographical Surveys, Appendix JJ., dated 1876, cited as Henshaw, 1876. (Observations at Santa Barbara, Fort Tejon, Mount Whitney, Walker's Basin and intervening country in 1875.)

H. W. Henshaw. Wheeler's Surveys, 1879, Appendix OO., in this report quoted as Henshaw, 1879. (Mr. Henshaw's observations were on the east slope from Carson to the Columbia River, and included work in the summers of 1877 and 1878.)

O. B. Johnson. Birds of Willamette Valley. American Naturalist, 1880. Prof. Johnson also contributed his observations at Seattle in 1884.

Capt. Charles E. Bendire, U. S. A. Notes on some of the birds found in southeastern Oregon, particularly in the vicinity of Camp Harney. Proc. Boston Society of Nat. Hist., 1877. (Observations from November, 1874, to January, 1877.)

Robert Ridgway. United States Geological Survey

of the Fortieth Parallel. Part 3. Ornithology. Prof. Ridgway made observations at Sacramento from June 6th to July 4th, 1867, afterward crossing the Sierra to Truckee Valley, Pyramid Lake, Carson and other localities in western Nevada, where the remainder of 1867 was spent. Observations were continued in western Nevada until July, 1868. Afterward the route from Austin to Salt Lake was explored.

Dr. W. J. Hoffman. Annotated List of the Birds of Nevada, Author's Edition, Vol. 6, U. S. Geological and Geographical Survey. Dr. Hoffman followed the 115th meridian from near northern Nevada, southward to about 37° latitude, thence westward to Owen's Valley, California, and from this point southeast to Fort Mojave, and thence up the Colorado River beyond Nevada.

Baird, Brewer and Ridgway, North American Birds. No doubt the most complete work on North American birds ever published in America. It has also a merit that is getting rarer of late, that of giving credit to whom credit is due.

Bulletin of the Nuttall Club. A quarterly journal of ornithology. The first volume was issued in 1876. It afterward became the Auk.

The Auk, the organ of the American Ornithologists' Union.

In addition to the forenamed authorities, a few items have been culled from other publications. The Proceedings of the Phila. Acad., containing Dr. Gambel's papers, and the Proc. Cal. Acad. Sci., containing some of Dr. Cooper's articles, came into my possession too late to benefit much by them.

There being no correspondents in Nevada, I have drawn freely upon the valuable reports and papers of Messrs. Bendire, Henshaw, Ridgway and Hoffman.

By Central California I refer to the part of the State

between the northern parts of Stanislaus and Tuolumne counties and the northern part of Butte, southwestern Plumas and Sierra counties.

I have made observations at many localities in this part of the State, in the tule swamps, river bottoms, plains, foothills and coniferous forests of the Sierra Nevada Mountains at all altitudes, kept a record of the birds, but have not thought it necessary to burden my notes with a long list of localities. The summer residents are the same in northern Tuolumne County as in northern Butte, though a few species become more numerous with increase of latitude, and there is a corresponding decrease in altitude of the breeding range of some of the mountain species. There is little difference in the resident species of the northern Sacramento Valley and the southern San Joaquin Valley, and I believe the avifauna of Central California nearly represents that of the State north of about the 35° of latitude, east of the coast mountains and west of the Great Basin, though a considerable portion of this tract has not been ornithologically explored, the Sierra from near Tehachapi to Alpine county having been quite neglected.

I am quite confident that few, if any, species have escaped my notice in Central California, except a few which probably visit the high Sierra Nevada in winter, from the north, when snow is so deep as to prevent exploration.

The Pacific District has an area, exclusive of British Columbia, of 434,000 square miles. California alone is more than twice as large as the six New England States, has a great diversity of surface and climate, and is as long as from Florida to Lake Erie—facts sufficient to prohibit positive opinions until after a more thorough exploration.

The nomenclature is that at present sanctioned by the

American Ornithologists' Union, which has changed considerably since this paper was commenced. Mr. Walter E. Bryant is entitled to credit for making the necessary changes and otherwise preparing it for the printer.

1. Colinus virginianus (Linn.) Bob-white.

Mr. Ramon E. Wilson, of the California Sportsman's Association kindly furnishes the following concerning introduced game birds; date, October 12, 1885: "Our efforts in that direction have resulted in failures, except as I will state. Mr. Estee, some years ago, placed two dozen bob-whites on his farm in Napa County. Every precaution was taken to protect them from hunters, and they were carefully looked after. They all soon disappeared, the theory being that they were destroyed by vermin. I learn that last February some of the same kind of birds were placed on the farms of Mr. Miller, Mr. Samuel Rea and Mr. J. P. Sargent, along Carnedero Creek, near Gilroy. It is said they have bred the past season, and their numbers materially increased. The experiments, however, from the length of time can hardly be called a success.

Some years ago some Arizona quail were put out near Folsom, but they all soon disappeared, and nothing has been heard of them since. Some bob-whites were placed on General Bidwell's place near Chico, but I understand they have disappeared also. Bob-whites roost on the ground and are therefore unable to protect themselves from the vermin which is so plentiful everywhere in California.

Some years ago a flock of English pheasants was put out in the woods of Santa Cruz county, but nothing has been seen nor heard of them since. Colonel Haymond of San Mateo has a number of these birds, English and

Japanese, but he has had no success in raising them; when let out they suddenly disappear and nothing is seen or heard of them. Mr. Howard, near by, has experimented with the same bird. A few weeks since he informed me that his foreman told him he had seen a flock of twenty-two birds. The birds mentioned are the only ones experimented with. Certainly thus far the experiments are not a success. In Oregon they have met with great success, with both quail and pheasants."

Bendire (1877). This species may properly be included in the avi-fauna of southern Oregon. It was originally introduced at Boise City, Idaho, and now extends to the Oregon side of Snake River, and is multiplying rapidly.

Willamette Valley. O. B. Johnson (1880).—Introduced and doing finely.

2. **Oreortyx pictus** (Dougl.) MOUNTAIN PARTRIDGE.

Newberry. They extend from the Columbia almost uninterruptedly, but nowhere abundantly through the Siskiyou, Calapooya and Trinity mountains to California.

O. B. Johnson. Very common throughout Western Oregon, breeding extensively.

Cooper, 1860. Very rare in Washington Territory, a few small coveys only being found about Vancouver as I was informed by the officers of the garrison in 1853.

3. **Oreortyx pictus plumiferus** (Gould). PLUMED PARTRIDGE.

San Diego County. F. E. Blaisdell.—Common in the higher mountains.

Volcan Mountains. W. O. Emerson.—Observed January 23, the day of my arrival, and only once afterward; probably the snow drove them down the mountain.

Henshaw, 1876. We found it in the mountains near Fort Tejon and in the Sierra in sufficient number of localities to justify the belief that its distribution in southern California is quite general.

Agua Caliente, San Diego County. F. Stephens.—Resident; heard almost daily in the foothills.

San Bernardino. F. Stephens.—Tolerably common in the mountains; breeds.

Tehachapi. L. B.—Common resident.

Paraiso Springs, Monterey County. W. E. Bryant.—April 1885.

Baird, Brewer and Ridgway. An egg of this species taken by Dr. Canfield near Monterey.

Central California. L. B.—Very common throughout the Sierra in summer and is increasing rather than otherwise. Probably a few nests are destroyed by sheep, and a few are deserted in consequence of sheep grazing about them, but this quail does not desert her nest for slight cause. Many cross from California to the east slope to breed, and, having done so, in September they begin to cross to the west slope where they winter at varying heights according to season. Between Summit and Donner Lake, as early as September 4, 1885, I met several flocks coming up from the lake on their way down the west slope. The most of the species had passed west of Summit before October 1, although I found a few as late as October 12, and a few winter on the east slope, as the settlers of Sierra Valley told me this was the case there, and Mr. Ridgway found a flock near Pyramid Lake, December 27, 1867. One of his specimens was taken at Carson City, March 10, 1868. They frequently take shelter in the snow sheds of the Central Pacific Railroad during early snow storms and sometimes journey westward within them. At such times they often appear stupid and appear to have lost the instinct of self-preservation but it may be that they become dazzled by the snow if not quite snow-blind. Their nests are usually in rock rose, also called "tar weed" *(Chamæbatia foliolosa)*, or in "bear bush," "snow bush" *(Ceanothus cordulatus)*. One I saw was in a hollow stump, but, as usual, was on the ground and was probably made when the snow was several inches deep outside of the stump. The eggs appear to be in most cases from eleven to thirteen. The first broods are out from about June 20 to July 5, according as the season is early or late.

Cisco. Dr. Cooper (in letter). Cisco, altitude 5,911 feet, April 28, on snow.

Igo. E. L. Ballou.—Common resident. March 27, 1884, the species was mated here. June 1, sitting.

Henshaw, 1879. It is only at rare intervals that it appears to cross the mountains and appear along the eastern slope. About Carson, at Eagle and Honey Lakes, California, and at the Dalles on the Columbia, their presence was detected and is to be accounted for through the natural dispersion of the species. At several localities near Camp Bidwell, Cal., the several covies are the descendants brought from the Pacific Slope and let loose to shift for themselves. They are nowhere in this region very numerous.

Vancouver and adjacent islands. John Fannin— Introduced from California.

[Perhaps those seen at the Dalles by Mr. Henshaw belonged under *O. pictus*. Those I found in Butte Co., lat. 40° 10′, altitude about 5,000 feet, were identical with those found farther south in the Sierra Nevada. I believe San Pedro Mountain, Lower California, is the most southern known range of the genus *Oreortyx*. I am positive that it does not inhabit Lower California south of La Paz, and doubt if it ever did. San Pedro Mountain, as it is termed by the people of northern Lower

California, is about 125 English miles south of the boundary line, and on the charts of the U. S. Coast Survey bears the name of Santa Catalina Mountain, and also that of Calamajuet, which latter name I have sometimes applied to San Pedro.]

4. **Callipepla californica** (Shaw).   CALIFORNIA PARTRIDGE.

This, the typical form of the California quail or partridge, inhabits the coast of northwest California, Oregon and Washington. It is darker above than the interior and southern form of the "valley quail," as all of these birds are universally termed in California. Sportsmen know but two kinds of quails indigenous to this coast, namely valley quails and mountain quails, and these names are likely to be perpetuated.

5. **Callipepla californica vallicola** Ridgw.   VALLEY PARTRIDGE.

Abundant about San Diego, where, March 7, 1884, the breeding note of the male was first heard; March 27th scattering in pairs in cañons. April 4, a pair excavating for a nest under a small cactus. May 13, Mr. W. J. Morgan saw a brood just hatched, the first of the season as far as known; spring wet and cool. Tia Juana Valley near San Diego, April 3, 1885, two broods of young about a week old. Season unusually early.

Poway. F. C. Blaisdell.—An abundant resident; first set of eggs taken April 27, 1884.

Volcan Mountains, altitude about 5,000 feet. W. O. Emerson.—Large flocks all winter.

Aqua Caliente, San Diego Co., Cal. F. Stephens.—March 18 to April 15, 1886. In small numbers in the outer edge of the desert along foothills.

San Bernardino. F. Stephens.—An abundant resident. Santa Catalina Island; probably introduced.

Tehachapi and vicinity. L. B.—Abundant resident; *C. gambeli* not seen by me.

Henshaw, 1876–'79. Near Fort Tejon I saw the species on several occasions at an altitude of 6,000 feet. Its numbers in some sections of the State are simply enormous. (Nowhere indigenous along the eastern slope.)

Dr. Cooper, 1870. The only point east of the Sierra Nevada where I have seen this bird is along the upper part of the Mojave river.

Alameda and Contra Costa counties. W. E. Bryant.—Abundant resident.

Berkeley. T. S. Palmer.—Common resident.

Central California. L. B.—From common to abundant in suitable localities, in summer, up to 5,000 feet altitude; chiefly about dwellings at this altitude. Rather rare at Red Bluff where much of the country is used for pasturing sheep; formerly very abundant in the Marysville Buttes but now rare for the same reason. Not only do sheep destroy nests by treading on them but they prevent the growth of cover, and this very timid bird deserts her nest where there is the least cause for doing so.

Chico. Wm. Proud.—In General Bidwell's park and orchard, June 6, 1884, first young; next brood on June 12; June 15 coming out plentifully. On June 14 I counted eight males within a radius of less than one hundred yards and not a female among them, they still being on the nests. I recently found a nest containing twenty-two eggs, the bird sitting. I think this the largest number I have yet found in one nest. The valley quail will, under some circumstances, breed twice but it is seldom they raise two broods. April 1, 1885, first nest, nine eggs; April 15, first young, probably two days old. April 17, sharp frost, snow on the mountains.

Igo. E. L. Ballou.—July 1, 1884, first young, not large enough to fly but large enough to hide.

Oakland, Oregon. W. E. Bryant.—Seen in breeding season.

Calaveras County foothills, altitude 1,000 feet, June 7, 1885. Jas. Brice.—A flock of young about the size of a chicken just from the egg; I think they could not fly; the first I have seen on my route this year.

Mr. Brice travelled daily from Murphy's to Milton in the hills where this quail is more than abundant, and as I had passed over it a few days before without seeing any I requested him to report the first young, which he did as above.—L. B.

Poway. F. E. Blaisdell.—First young seen May 5, 1884. They bred very abundantly until late in August. The last set of fresh eggs were taken August 14, 1884. I think it was unusually late for them to lay. I also noticed some small quails about November 1.

6. Callipepla gambeli (Nuttall). GAMBEL'S PARTRIDGE.

Agua Caliente, San Diego Co. F. Stephens.—In the Colorado desert, March 25 to 28, several seen, one shot.

March 18 to April 15, 1886. Not so common as *C. c. vallicola* in the same locality. Are inclined to keep further in the desert.

In San Bernardino County, along the line of the Southern Pacific Railroad, *L. californicus* and *L. gambeli* come together; here hybrids occur. (Henshaw, Auk., July, 1885, referring to specimens shot near San Gorgonia Pass, by R. B. Herron.

Cooper, 1885. Their range toward the north is not known to be above 36°. At Fort Mojave they are numerous and have all the calls of the coast species except the alarm chirp like a robin's which I never heard them utter. There is however a slight difference in their notes, which is recognizable by strangers.

Heermann. I first discovered this beautiful species on

the Mojave Desert where the Mojave River empties into a large salt lake, forming its terminus. At Fort Yuma they were quite abundant, congregating in large coveys.

[I have found this bird very wild, and again very tame, owing no doubt to its experience, or inexperience, with mankind.]

7. **Dendragapus obscurus fuliginosus** Ridgway. SOOTY GROUSE.

Henshaw. In California it is found in both the coast and Sierra ranges as far south as latitude 35° and probably even lower. It was present though not very common in the mountains near Fort Tejon and was rather numerous in the region about Mount Whitney.

L. B.—Calaveras county, altitude about 3,500 feet to summit of Sierras, not common, mostly in rugged localities. I think but few of their nests are destroyed by sheep and that they usually hatch before the mountains are overrun by the large droves of sheep which are annually driven from the valleys to the mountains to spend the summer and spread desolation and dust everywhere. The following will give some idea of the time of hatching: Calaveras Big Trees, June 25, 1881, female parent and six or seven young, the latter about a week old; season early. Big Trees, June 14, 1882, female adult and chicks. Summit, 7,000 feet altitude, July 4, 1885, a brood of young about two weeks old; a few old males still hooting or grunting, but the most of them had abandoned the society of the hens and gone into the high peaks. I found as early as August 5, young nearly two-thirds grown.

I could not find any of these birds after about October 1st, when they were probably in the evergreen trees, their usual winter quarters. I have never seen more than seven young in a brood, though perhaps some escaped my notice, as Captain Bendire says the full complement of the Camp Harney bird is from eight to ten.

Newberry. Found not uncommonly in the Sierra Nevada in California and in the wooded districts lying between the Sacramento Valley and the Columbia.

Willamette Valley. O. B. Johnson.—Common resident, breeding extensively.

Cooper, 1860. Common in most of the forests of the Territory.

Suckley, 1860. It is common on the east side of the Cascades as far north as the 49th parallel.

British Columbia. John Fannin.—An abundant resident.

Henshaw, 1879. The blue grouse, which is found in the Sierra and Cascade ranges, at least along their eastern slopes and as high up as the Columbia River, is the typical middle region form. The mountain forests, especially those composed largely of firs and spruces, abound with this fine game bird. Several broods of young chicks were found about the middle of June.

Camp Harney. Bendire, 1877.—A common resident throughout the mountains. We have two varieties, *fuliginosus* and *richardsoni*, the former being the most abundant. In winter they seldom alight on the ground, excepting to get water.

Ridgway. More or less common on all the ranges clothed with a sufficient extent of pine forest. It was found on the Sierra Nevada, near Carson, and on several of the higher ranges of the Great Basin.

8. Dendragapus obscurus richardsonii (Sab.) RICHARDSON'S GROUSE.

Camp Harney. Bendire.—Less abundant than var. *fuliginosus*.

9. **Dendragapus franklinii** (Dougl.) FRANKLIN'S GROUSE.

British Columbia. John Fannin.—An abundant resident east of the Cascades.

Suckley, 1860. Abundant in the Rocky and Bitter Root Mountains, also found in the Cascades in the Yakima passes.

10. **Bonasa umbellus umbelloides** (Dougl.) GRAY RUFFED GROUSE.

Camp Harney. Bendire.—Rare resident about here, frequenting densest undergrowth along the mountain streams, and seldom seen.

11. **Bonasa umbellus sabini** (Dougl.) OREGON RUFFED GROUSE.

Wilbur, Oregon. W. E. Bryant.—Summer of 1883, young and old shot.

Willamette Valley. O. B. Johnson.—Very common along water-courses, where it breeds. Seattle, May 1st, 1884, nest and fresh eggs.

Cooper, 1860. Very abundant everywhere about the borders of woods and clearings; it is common near the forest east of the Cascade Mountains.

Suckley, 1860. Abundant in the timbered districts throughout Washington and Oregon. In habits they are identical with the same bird east. Owing to the mildness of the season in the vicinity of Fort Steilacoom, the males commence drumming as early as January, and in February I have heard them drumming through the whole night.

British Columbia. John Fannin.—An abundant resident.

Henshaw, 1879. This form of the ruffed grouse occurs abundantly along the eastern slope, although perhaps not until Oregon is entered (from the south). Fort

Klamath was the first point at which I obtained undeniable proof of its presence. The grouse of this region, while referable as above, do not typically represent the variety *sabini*, which reaches its maximum of development, as indicated by depth of color and redness of tint, only on the Pacific slope.

Camp Harney. Bendire.—Moderately common in the John Day River Valley.

12. **Lagopus rupestris** (Gmel.) ROCK PTARMIGAN.

British Columbia. John Fannin.—Common resident.

13. **Lagopus leucurus** Swains. WHITE-TAILED PTARMIGAN.

British Columbia. John Fannin.—Common resident east of the Cascades.

14. **Pediocætes phasianellus columbianus** (Ord.) COLUMBIAN SHARP-TAILED GROUSE.

Newberry.—Found as far west and south as the valleys of California. We first found it on a beautiful prairie near Canoe Creek, about fifty miles northeast of Fort Reading. Subsequently, after passing the mountain chain which forms the upper cañon of Pit River we came into a level grass-covered plain. On this plain were so many that they afforded us fine sport and an abundance of excellent food. We found them again about the Klamath Lakes and in the Des Chutes Basin quite down to the Dalles.

Fort Klamath. Lieut. Wittich, 1879.

Henshaw, 1879. Appears to be entirely absent from eastern California and western Nevada except in the upper districts. About Camp Bidwell the "sharp tails" are sufficiently numerous to afford excellent shooting. Farther north in Oregon, and especially on the grassy plateaus that border the Columbia River, and on the rolling hills for a hundred miles south, it is extremely abundant.

Camp Harney. Bendire.—Only a moderately common resident, apparently irregularly distributed. In the vicinity of Camp Harney they are mostly found in the juniper groves during cold weather, and the birds live almost exclusively on the berries of these trees. The eggs usually number from eleven to fourteen.

Hoffman. Found in moderate numbers at Bull Run Mountain.

Ridgway. This grouse, known universally among the western people as the "prairie chicken," we found in the upper Humboldt Valley near "Trout Creek," where it was abundant.

British Columbia. John Fannin.—Tolerably abundant; only east of the Cascades.

[Perhaps Mr. Fannin's note refers to the northern sharp-tail *Pediocætes phasianellus*, the habitat of which is given in the recent check list of the A. O. U. as follows: British America, from the northern shore of Lake Superior and British Columbia to Hudson's Bay Territory and Alaska.]

15. Centrocercus urophasianus (Bonap.) SAGE GROUSE.

Cooper, 1860. Common on the high barren hills and deserts east of the Cascades.

Suckley, 1860. Abundant on the sage plains of Oregon near Snake River on both sides of the Blue Mountains. They are also found along the line of the Columbia River on the open plains and again on the sage barrens of the Yakima and Simcoe valleys in Washington Territory about latitude 46°–47° north, in fact wherever "sage" (*Artemisia*) abounds this bird is found.

Camp Harney. Bendire.—A common resident species, particularly abundant in the upper Sylvies Valley at an altitude of about 6,000 feet.

Henshaw, 1879. Numerous as is this species in many portions of the Rocky Mountain region it appears to be even more abundant in the sterile tracts that lie just east of the Sierra Nevada and Cascade ranges.

Ridgway. Abundant in certain localities but by no means evenly distributed.

Hoffman. The only locality where this bird was found at all common was near Belmont. Specimens were also shot at Hot Creek Cañon.

L. B.—Common resident of Sierra Valley and occurs in Alpine county up to 8,500 feet but rare so high; also found in Mono county by Lieutenant McComb.

Newberry. On the shores of Wright and Rhett lakes we found them very abundant.

16. Columba fasciata Say. BAND-TAILED PIGEON.

I saw a flock in El Cajon, San Diego county, December 15, 1883, a rare occurrence.—L. B.

Poway. F. E. Blaisdell.—I have seen this species here on three occasions. It was very abundant in the Volcan Mountains in September and October when choke cherries were ripe.

Henshaw, 1876. Not seen by us until in the fall. In November I often saw them in flocks of from ten to one hundred.

Cooper, 1870. North of San Francisco I have seen them in flocks as early as July and at the Columbia River they spend the summer in the valleys as well as throughout the mountains. They are there migratory, leaving in October, but in California their wanderings are guided chiefly by want of food. I have found them building in the Coast Range as far south as Santa Cruz.

L. B.—It is rare in the mountains of Central California.

in summer, but usually quite common in the foothills in winter. I have shot them in July, in Calaveras County, with their crops full of pine nuts.

Igo. E. L. Ballou.—A migrant and resident. On March 26 and 27, 1884, a flock seen; also during our heavy snow storms, five weeks previously, from which I infer that the storms and pigeons had some connection.

Cape Foulweather. S. L. Wass.—Resident.

Willamette Valley. O. B. Johnson.—An abundant summer resident.

Cooper, 1860. Arrives at Columbia River in April.

Suckley, 1860. I saw but one flock containing five individuals, east of the Cascade Mountains.

British Columbia. John Fannin.—Tolerably common summer resident.

San Jose. A. L. Parkhurst.—April 19, 1885, large flocks, the last seen.

Beaverton, Or. A. W. Anthony.—March 29, 1885, first seen (ten birds); next seen March 30; April 21 common. Common in breeding season.

Admirality Head, Whidby Island, W. T. Lawrence Wessel.—April 22, 1885, first seen.

Burrard Inlet, B. C. John Fannin.—April 5, 1885, first seen (two males); next seen April 12; May 4 common. Common in breeding season.

17. **Ectopistes migratorius** (Linn.) PASSENGER PIGEON.

Ridgway. Only a stray individual was met with by us, and it cannot be considered as more than an occasional straggler in the country west of the Rocky Mountains.

Mr. Ridgway's specimen was shot at West Humboldt Mountains, Nevada, September 10, 1867.—L. B.

18. **Zenaidura macroura** (Linn.) MOURNING DOVE.

San Diego. L. B.—Common in winter; abundant in summer. Very common in the northern 100 miles of Lower California in May, and breeding. At San Diego, spring of 1885, the first nest I found was on April 17; incubation far advanced. They raise several broods in a season, and I have seen eggs in Calaveras County in the first part of September.

Poway. F. E. Blaisdell.—Common in winter, abundant in summer. The first eggs seen in 1884 were taken April 14.

Santa Isabel. W. O. Emerson.—Wintered here.

San Bernardino. F. Stephens.—Breeds abundantly in the valleys.

Agua Caliente. F. Stephens.—March 25-28, not common. March 18 to April 15, 1886. First seen March 27; common after April 1. Summer resident.

Henshaw, 1876. Very numerous in southern California.

Santa Cruz. Jos. Skirm.—First seen April 10, 1882; April 6, 1883.

Alameda and Contra Costa counties. W. E. Bryant. Abundant summer resident.

Haywards. W. O. Emerson.—Common summer resident. April 23, first; next seen May 1, 1885.

Berkeley. T. S. Palmer.—Tolerably common summer resident. December 4, 1884, tolerably common; never knew it to remain so late. First seen April 30, 1885.

Nicasio. C. A. Allen.—First seen April 20, 1876; April 30, 1884.

Olema. A. M. Ingersoll.—First seen April 18, 1884.

Central California. L. B.—Abundant summer resident in valleys and foothills. Seen at Red Bluff February 3, 1885, but rare; Chico, February 6-7, rather common, and cooing. Common in flocks throughout ordinary winters, if not every winter, as far north as Yuba and Butte counties.

Chico. Wm. Proud.—February 27, 1885, one specimen.

Beaverton, Or. A. W. Anthony.—Common summer resident. First seen April 29, 1884; rare until about June 1st; abundant June 7. First seen March 30th, 2 specimens; common April 10, 1885.

Willamette Valley. O. B. Johnson.—An abundant summer resident.

Walla Walla, W. T. Dr. Williams.—April 24, four; common May 5, 1885.

Suckley, 1860. Very abundant throughout Oregon and Washington Territories.

British Columbia. John Fannin.—Rare summer resident. Burrard Inlet. First seen May 7; common May 20, 1885.

Carson. Henshaw, 1879. Extremely numerous, not only here but all along the eastern slope far up into Oregon and Washington territories.

Camp Harney. Bendire.—An abundant summer resident, arriving about May 1. They rear but a single brood in a season, while in Arizona I found fresh eggs as late as September 14.

Hoffman. Generally distributed over the whole State.

Dr. Cooper. Truckee, April 29, 1870.

Carson. Ridgway. Arrived April 23, 1868.

19. **Melopelia leucoptera** (Linn.) WHITE-WINGED DOVE.

F. Stephens (Bull. Nutt. Orn. Club, January, 1883). At Yuma they were actually common, but none were found to the westward of this point.

They are abundant in the Cape St. Lucas region and probably inhabit the coast on the west side of the Gulf of California from Cape St. Lucas to Fort Yuma.—L. B.

20. **Columbigallina passerina** (Linn.) GROUND DOVE.

Baird, Brewer and Ridgway, Vol. 3, 522. Dr. Cooper states that an individual of this species was killed by Mr. Lorquin at San Francisco in May, 1870. Mr. Lorquin also obtained several at San Gabriel, Los Angeles County, several years previous.

21. **Pseudogryphus californianus** (Shaw). CALIFORNIA VULTURE.

San Diego. L. B.—Generally reported to be a resident of the mountains in this part of the State, but not seen here or in any part of Lower California by me, though Col. N. S. Goss informed me that one or more pairs breed near Mr. Crosswaith's ranch about 60 miles south of San Diego. I have not seen one of these birds in the field in ten years. I was told at Tehachapi, in the spring of 1889, that a few still breed between Tehachapi and Tejon Valley.

Poway. F. E. Blaisdell.—A rare species in this region; occasionally seen on Volcan Mountains from August 21 to November 28.

San Bernardino. F. Stephens.—Very rare resident of the valley and mountains.

Henshaw, 1876. Our opportunities for an acquaintance with this vulture were limited to seeing two or three individuals.

Santa Cruz. Joseph Skirm.—Tolerably common. I have seen them in a flock in company with *Cathartes aura*. It journeys along the coast.

Chico. Wm. Proud.—Sometimes seen.

Cooper, 1870. I have not seen many of these birds along the coast where most of my later collections were made, and none on the islands or in the highest Sierra Nevada.

Newberry, 1855. This vulture, though common in

California, is much more shy and difficult to shoot than its associate, the turkey buzzard, and is never seen in such numbers. We saw very few in the Klamath Basin and none in Oregon.

Cooper, 1870. This confirms the observations of Dr. Suckley and myself, as we saw none during a long residence and travels near the Columbia, except one which I supposed to be this, seen at Fort Vancouver in January. Like several other birds seen there by Townsend and Nuttall, they seem to have retired more to the south since 1834.

British Columbia. John Fannin.—Very rare summer resident.

Nuttall, 1840. According to Douglass in the Zoölogical Journal it is common in the wooded districts, migrating in summer as far north as the forty-ninth parallel.

[It is difficult to believe that this was ever really an abundant species in California. It has certainly been very rare in the center of the State north of latitude 38° since the spring of 1856. Its present rarity may be accounted for by quoting Dr. Cooper (Cal. Orn.): "It is often killed by feeding on animals such as bears, when poisoned by strychnine by the rancheros; the poisoned meat kills them readily. The rancheros have very little fear of their depredations on young cattle, though it has come within my knowledge for five or six to attack a young calf, separate from its mother, and kill it; the Californians also say they are often known to kill lambs, hares and rabbits."]

Heermann, 1854. This bird was observed occasionally during our survey. Whilst hunting unsuccessfully in the Tejon Valley we have often passed hours without one of this species being in sight, but on bringing down any large game, ere the body had grown cold these birds might be seen rising above the horizon, slowly sweeping towards us, intent upon their share of the prey. Nor, in the absence of the hunter, will his game be exempt from their ravenous appetite, though it might be carefully hidden and covered by shrubbery and heavy branches. I have known these marauders to drag forth from its concealment and devour a deer within an hour.

Gambel (Phila. Acad.) Particularly abundant in winter, when they probably come from Oregon.

22. **Cathartes aura** (Linn.) TURKEY VULTURE.

San Diego. L. B.—Common resident; common in northern and southern Lower California, and probably in all parts of it. This bird, the caracara eagle and other very useful carrion-eating birds are wisely protected by law in Mexico, as should be the case with this in our country, being harmless in all respects and very useful as scavengers.

Poway. F. E. Blaisdell.—Common resident.

Volcan Mountains. W. O. Emerson.—February 9, five observed.

Contra Costa County. W. E. Bryant.—Common resident; breeds.

Central California. L. B.—Very common constant resident below the pine forest; often seen in the high Sierra in summer; no doubt a common species throughout the State. One that I shot on a ranch at Gridley appeared to be catching grasshoppers, but as it had recently feasted on dead colt it was too offensive for thorough inspection. There were many on this ranch, and if they did not partake of the grasshoppers, or, rather, of the true middle province locust, they were about the only exception besides the turtle dove, as all the other birds from the large red-tailed hawk and crow down to the little plain titmouse fairly feasted on the destructive insect; even the egg-sucking magpies and blue jays became more than useful, and this continued throughout the most of the summer.

Willamette Valley. O. B. Johnson.—Common during summer.

Beaverton, Or. A. W. Anthony.—March 22, 1885, first seen; next seen April 15; rare.

Cooper, 1860. Very abundant in all parts of the Territory I have visited. They arrived at Puget Sound about the middle of May and flocks could be seen daily about the carcasses of sheep.

British Columbia. John Fannin.—Tolerably common.

Henshaw, 1879. Generally distributed. In some localities as near Honey Lake, California, very numerous.

Camp Harney. Bendire.— Moderately common during the summer months and breeding in this vicinity. They arrive here early in April and I saw one on the 27th of November after a fall of snow.

Hoffman. Arrives in the middle regions about the second or third week in March, after which time it was was common. Was also observed in the Colorado Valley from Fort Mojave northward.

Agua Caliente. F. Stephens.—March 25, rather common. Abundant March 18, 1886, and seen often after up to the day of leaving (April 15).

Ridgway. In the interior it was abundant throughout the summer when it was found in nearly all localities, but during the winter months they seemed to have retired to the southward, none having been seen in the latitude of Carson earlier than the middle of March.

23. **Elanus leucurus** (Vieill.) WHITE-TAILED KITE.

Ventura County. B. W. Evermann.—A rare resident. I obtained a full set of eggs April 12. (Auk, vol. iii, 1886.)

Baird, Brewer and Ridgway. Dr. Gambel, who secured his specimens at the mission of St. John, near Monterey, describes it as flying low and circling over the plains in the manner of a marsh hawk.

Alameda County. W. E. Bryant.—This hawk bred here formerly; it is still a very rare resident.

Sebastopol. F. H. Holmes.

Berkeley. T. S. Palmer.—Rare accidental visitant. Found only about the marsh, three miles west of here on the bay. December 26, 1885, again April 13, 1886.

Cooper, 1870. I have seen them as far south as Bolinas Bay and near Monterey, but always about streams or marshes.

Heermann. The extensive marshes of Suisun, Napa and Sacramento Valleys are the favorite resorts of these birds, more especially during the winter season, as they then find a plentiful supply of insects and mice, their principal nourishment. I fell in with an isolated pair in the mountains between Elizabeth Lake and Williamson's Pass.

L. B.—This hawk is still a common resident about the extensive tule marshes in the center of the State. I have seen what I believed were their old nests in willow trees along the San Joaquin River. I noticed one of these birds at Gridley, October 20, 1884, and another at Red Bluff, February 4, 1885. It is a common resident in the tules of Sutter County.

24. **Circus hudsonius** (Linn.) MARSH HAWK.

Northwest Lower California. Col. N. S. Goss.—Near the boundary line, March 23, 1884, nest and eggs.

Tolerably common about San Diego.—L. B.

Santa Ana River. F. E. Blaisdell.— December 9, 1884, to January 6, 1885, not common.

San Bernardino Valley. F. Stephens.— Tolerably common resident.

Agua Caliente, San Diego County. F. Stephens.— One was observed on each of the following days: March 18, April 3 and April 8, 1888.

Henshaw, 1876. Very numerous in California; resident in the southern part.

Cooper, 1870. One of the most abundant hawks throughout the unwooded country and about every marsh, even in the dense forest.

Alameda and Contra Costa counties. W. E. Bryant. Rare, resident (?)

Central California. L. B.—Common resident in the valleys; rare summer resident of the mountain meadows.

Willamette Valley. O. B. Johnson.—Moderately common; breeding.

Fort Klamath. Lieut. Wittich.—Abundant.

Cooper, 1860. Abundant throughout the open districts of the Territory, especially in winter, and it breeds there.

British Columbia. John Fannin.—A common summer resident.

Henshaw, 1879. Very numerous in every suitable locality.

Camp Harney. Bendire.—Moderately abundant and a few resident.

Ridgway. No marsh of any extent was visited, either in winter or in summer, where this hawk could not be seen at almost any time during the day skimming over the tules in search of its prey.

Newberry. Rather common in the Sacramento valley, and abundant beyond all parallel on the plains of Pit River.

25. Accipiter velox rufilatus Ridgw. WESTERN SHARP-SHINNED HAWK.

San Diego. L. B.—Tolerably common winter visitant.

Volcan Mountains. W. O. Emerson.—Seen February 22.

Volcan Mountains. F. E. Blaisdell.—From August 21 to November 28, not common.

Cooper, 1870. They probably breed more generally towards the southern and lower parts of the State than that species *(A. cooperi)*, as I have seen a few of them (or perhaps *A. mexicanus)*, in the warmer months.

Henshaw, 1876. A common resident throughout southern California.

L. B.—Common during summer months in the upper Sierra. (Birds Central Cal., 1879). My later observations are greatly at variance with the statement of 1879. During the summer of 1885, nearly all of which I spent at the Summit of the Central Pacific Railroad, altitude 7,000 feet, and upward, I saw but a single individual, a fine adult male, shot July 4. About September 15, it became common; still more so about October 1. It may be a common summer resident in a very few localities in central and southern California, especially in mountains having the height of Mount Whitney where Mr. Henshaw probably saw it.

Alameda and Contra Costa counties. W. E. Bryant.—Tolerably common winter visitant.

Haywards. W. O. Emerson.—March 21, 1884, last seen; first seen in fall, August 28.

Berkeley. T. S. Palmer.—March 9, 1885, last seen. Tolerably common winter visitant. In 1886, seen March 7 and April 12.

Willamette Valley. O. B. Johnson.—Moderately common nesting in trees.

Cooper, 1860. Not common, and observed only in the colder months.

Suckley, 1860. Quite common near Fort Steilacoom latter part of summer and early autumn; quite scarce during breeding season.

British Columbia. John Fannin.—Common summer resident.

Henshaw, 1879. A single specimen only, at the Columbia River in October.

Camp Harney. Bendire.—Rather rare at all seasons.

Ridgway. Was observed only in the upper Humboldt where it was common in September.

26. **Accipiter cooperi mexicanus** (Swains.) WESTERN COOPER'S HAWK.

L. B.—Tolerably common about San Diego in winter; probably breeding in El Cajon; April 26, a single pair.

Volcan Mountains. W. O. Emerson.—Seen February 28, March 17 and March 20, 1884.

Volcan Mountains. F. E. Blaisdell.—August 21 to November 28, common. (1884).

Agua Caliente, San Diego Co., Cal. F. Stephens.—One seen April 8, 1886.

San Bernardino Valley. F. Stephens.—Rare resident.

Henshaw, 1876. In summer it is not often seen in the lower districts, but will then be found to have retired to the mountains. In the fall there appears to be a very decided migration from the north.

Cooper, 1860. A common species during the winter months in all the wooded portions of the State.

Alameda and Contra Costa counties. W. E. Bryant.—Rare winter visitant.

Central California. L. B.—Tolerably common in winter in the valleys and foothills; a few breed in the pine forest of Calaveras county, altitude 4,500 feet and upward, and I found a family along Feather River, July 25, 1885, where they had probably been reared.

Willamette Valley. O. B. Johnson.—Occasionally seen.

Cooper, 1860. Very abundant in summer.

British Columbia. John Fannin.—An abundant summer resident.

Henshaw, 1879. Appears to be more numerous in this region than the sharp-shinned hawk.

Camp Harney. Bendire.—Rare; seldom seen.

Hoffman. Not uncommon.

Ridgway. This daring depredator was more or less common in all localities where small birds abounded, but it was far from numerous anywhere.

27. Accipiter atricapillus striatulus Ridgw. WESTERN GOSHAWK.

British Columbia. John Fannin.—Summer resident; not common.

Suckley. I obtained several specimens both at Fort Dalles and Fort Steilacoom.

Cooper, 1860. It would seem to be the special frequenter of dark woods, where other hawks are rarely seen.

Henshaw, 1879. Seen at several points along the Cascade Mountains in Oregon.

Bendire. A few pairs breed in the mountains north of Camp Harney. Collected at Walla Walla.

Calaveras County. L. B.—Common summer resident; a pair or two breeding in the grove every summer; never positively identified in the foothills and valleys where it must be very rare if it occurs at all. It is occasionally found on the east slope of the Sierras in summer, as at Webber Lake, Donner Lake and Lake Tahoe. At the latter locality I saw about twenty at different times in September, 1889. It is usually in the densest forest.

28. **Buteo borealis calurus** (Cass.) WESTERN RED-TAILED HAWK. (*B. montanus* in part of Cooper and previous writers.)

San Diego, L. B.—Common resident.

Poway. F. E. Blaisdell.—Common resident.

Volcan Mountains. W. O. Emerson.—Seen on every collecting trip; paired and breeding February 20; eggs collected at Live Oak Spring March 20.

San Bernardino. F. Stephens.—Tolerably common resident.

Agua Caliente, San Diego county, Cal. F. Stephens. One seen on March 20, and another April 10, 1886.

Henshaw, 1876. Of almost universal distribution in the West.

San Jose. A. L. Parkhurst.—March 1, 1885, fresh eggs.

Alameda and Contra Costa counties. W. E. Bryant.—Common resident.

Berkeley. T. S. Palmer.—Common resident.

Central California. L. B.—Very common resident throughout the agricultural districts; many breed in the Sierra.

Newberry. It may be said to inhabit all portions of our Pacific possessions.

Willamette Valley. O. B. Johnson.—Common.

Cooper, 1860. An abundant and resident species in every part of the territory I have visited.

Suckley. I noticed that the poultry yards were as much harassed by this hawk as by the goshawk, neither of which hesitated to seize poultry from the very doors of dwelling houses. This may be accounted for by the fact that on Puget Sound hawks, as a class, are more numerous than in the Eastern States, and small birds to support them are less so in proportion.

British Columbia. John Fannin.—Common summer resident.

Henshaw. Numerous throughout all the heavily wooded portions of the region.

Camp Harney. Capt. Bendire.—Moderately common resident and one of the earliest hawks to breed.

Ridgway. A very common species in all wooded localities of the interior. Truckee and Carson Valleys, especially abundant in winter.

Hoffman. A common species throughout the wooded valleys in winter; ascending to the more elevated timbered regions in summer.

Buteo cooperi Cass. Cooper's Henhawk.—Has been placed in the hypothetical list of the A. O. U., 1886, with the remark that it is "probably the light phase of *B. harlani* Aud." (See discussions in the Auk, Vol. I, p. 253, and II, p. 165, on relationship of *B. harlani* with Cooper's henhawk.)

Cooper, 1870. The type specimen I shot near Mountain View, Santa Clara Valley, California, November, 1855.

29. **Buteo lineatus elegans** (Cass.) RED-BELLIED HAWK.

Willamette Valley. O. B. Johnson.—A single example referable to this species.

Heermann. Abundant, and extending from northern California to the edge of the Colorado desert.

Upper Sacramento Valley. L. B.—Apparently rare; probably resident, though not seen by me later than October 20. It is very common about Stockton in summer; nearly as common, in breeding season, as the red-tailed hawk. I knew of a pair nesting within less than two hundred yards of a residence where poultry was plentiful and easily obtainable. They nested there three consecutive seasons unmolested by the occupants of the dwelling. I shot the female as she flew from the nest, April 4, 1880. Her stomach contained several small lizards, a tree-frog *(Hyla)*, grubs and insects. May 11, 1879, there were three young in the nest that would weigh about a pound each. Mr. Charles Moore, who climbed the large oak in which the nest was placed, reported a lining of green but dried and broken leaves in the nest, about three inches deep in the center. On April 4, 1880, there were three nearly fresh eggs in the nest, which this year had a lining of the lace-like lichen *(Ramalina retiformis)*, found on the oaks in the vicinity, a sample of which was brought down from the nest by Mr. George Ashley, who, with great difficulty, secured the eggs. The largest egg measured 225 x 175; ground color of all bluish-white, much blotched with reddish-brown, the brown varying greatly in intensity. One egg was smaller and paler than the others, and was probably unfruitful. I saw one of these hawks at Stockton, January 25, 1885, repairing an old nest. This, too, was near a farm-house where poultry was abundant, but I doubt if it often attacks poultry, though I have known them to catch small birds.

Newberry. Common in those parts of California and Oregon traversed by our party.

Ridgway. Seen only in the Sacramento valley, where it was rather common among the trees along the river.

Cooper, 1870. Common in the southern part of the State, especially near San Diego. I did not see any in the Colorado valley.

El Cajon, San Diego County. L. B.—April 26, 1884, a pair about a nest. I saw one in Santa Margarita cañon April 26, 1885, and I also saw two in San Rafael valley, 95 miles southeast of San Diego, May 12, 1885.

Marysville. W. F. Peacock.—November, 1885.

30. Buteo abbreviatus Caban. ZONE-TAILED HAWK.

Cooper, 1870. I shot the specimen above described, the first found within the United States, on the 23d of February, 1862, thirty miles north of San Diego, and five from the coast.

31. Buteo swainsoni Bonap. SWAINSON'S HAWK.

Cooper, 1870. I found this species pretty common near San Diego, March, 1862.

San Diego. L. B.—Rare in winter, breeding in El Cajon, April 26. Shot in Lower California May 10, 1885, 40 miles southeast of San Diego; appears to be rare south of San Diego.

Poway. F. E. Blaisdell.—Only three observed, and these before April 1, 1885.

San Bernardino. F. Stephens.—Breeds in the valleys; rare.

Henshaw, 1876. Pretty well distributed over the southern part of California, and is in certain localities very common. This was the case in the San Fernando Valley in July.

Techachapi. L. B.—Common. Two shot in March, 1889.

Heermann. I first remarked this species at Graysonville Ferry on the San Joaquin River, and continued to meet it occasionally until we had crossed Kern River.

W. E. Bryant. Several specimens taken at Grayson in spring of 1881.

Stockton. L. B.—October 8, 1883, twelve seen on the ground in a single field catching insects. I have seen the species in Butte county in June and September; I suppose they breed there.

Fort Klamath. Lieut. Wittich.

Jacksonville, Oregon. W. E. Bryant.—In breeding season.

Henshaw, 1879. Very numerous in summer in the low, partially-wooded country near the mountains. I noticed more of these hawks in northern California than farther south, but this increase in numbers may have been only apparent and due to their concentration as fall approached.

Camp Harney. Capt. Bendire.—A common summer resident, generally distributed.

Hoffman. Frequently seen throughout the valleys, and appears to be more common during the autumn than *B. calurus*. Has also been obtained in the Colorado valley from Fort Mojave northward.

Ridgway. One of the most abundant of the large hawks of the interior, but it seemed to be less common in winter than in summer.

32. Archibuteo lagopus sancti-johannis (Gmel.) AMERICAN ROUGH-LEGGED HAWK.

Cooper, 1870. This species is only a winter visitor in California, as far as I know, and I have not seen them south of Santa Clara valley, though I think some may breed in the high mountains, as they are seen at the Columbia River in July.

Alameda and Contra Costa counties. W. E. Bryant. A rare winter visitant; one specimen taken at Berkeley.

Central California. L. B.—Tolerably common winter visitant; it may breed in the mountains, but I have no evidence that it does so.

Cooper, 1860. In October I found a large number near the seacoast; some remained all winter, and I think a few build near the mouth of the Columbia, where I saw young birds in July.

British Columbia. John Fannin.—Summer resident; not common.

Henshaw, 1879. Common in fall in marshy localities.

Camp Harney. Capt. Bendire.—An irregular winter visitor; common at some seasons and entirely wanting in others.

Ridgway. This common species was observed nearly everywhere in the vicinity of the fertile valleys. It appears to be resident in western Nevada, for it was extremely abundant in July at the Truckee Meadows.

33. Archibuteo ferrugineus (Licht.) FERRUGINOUS ROUGH-LEG.

British Columbia. John Fannin.—Summer resident; not common.

Suckley, 1860. An adult specimen taken in December, 1854, at Fort Dalles, Oregon.

Camp Harney. Bendire. Rather rare but more common in the open country to the southward, particularly so near Camp McDermitt where it breeds.

Ridgway. This magnificent hawk was much less frequently seen than its relative, the common rough-leg.

L. B.—I have never seen this hawk in California in summer, but it is rather common in the treeless lower foothills of Central California in winter, spending much of the time on the ground. It appears to be very rare in the upper Sacramento Valley.

Cooper, 1870. I found it common in December at Martinez.

Alameda and Contra Costa counties. W. E. Bryant.—Rare winter visitant.

Henshaw, 1876. It does not appear to be at all common in southern California, but becomes numerous in fall.

Heermann. During the recent survey in the southern part of the State I found it very abundant, having seen on one occasion in the mountains about sixty miles from San Diego, five or six of these birds at the same moment. Large tracts in the southern portion of the State being totally destitute of trees, this bird alights on the ground, or, taking a position on some slightly elevated tuft of grass or stone, will sit patiently for hours waiting for its prey.

34. Aquila chrysaëtos (Linn.) GOLDEN EAGLE.

Volcan Mountains. F. E. Blaisdell.—August 21 to November 28, occasionally seen.

Temecula Cañon. H. Willey.—December 2, 1881, one specimen.

Santa Cruz. Joseph Skirm.—Quite common resident.

Santa Catalina Island. F. Stephens.—August, 1886, four seen.

Alameda and Contra Costa counties. W. E. Bryant. Breeds; two records of capture.

Upper Sierras of Central California. L. B.—Rather rare summer resident; occasionally seen in the valleys in winter; formerly less rare; not seen at or south of San Diego by me.

Heermann. Rarely seen, save by the naturalist who is on the alert.

Ridgway. At Carson we scarcely ever went among the hills without seeing it soaring about, generally in pairs. We first met with it in July, 1867, near the summit of the western slope of the Sierra Nevada; afterward it was continually observed on all the higher ranges to the eastward.

Henshaw, 1879. Occurring more or less numerously among the mountains.

Camp Harney. Bendire.—Moderately abundant throughout the mountainous portions during the greater part of the year.

Hoffman. Not uncommon in the elevated mountain regions. They were observed at Bull Run Mountains, Mount Magruder and other similar localities.

35. **Haliæetus leucocephalus** (Linn.) BALD EAGLE.

British Columbia. John Fannin.—Very abundant resident.

Cooper, 1860. One of the most abundant of the Falcon tribe in Washington territory, particularly along the Columbia River and smaller streams, as well as salt water. This eagle is a constant resident.

Suckley, 1860. Exceedingly abundant in Oregon and Washington Territories, and in certain localities, especially during the salmon season, may be found in great numbers.

O. B. Johnson, 1880. Common along the Columbia River nesting in high trees.

Henshaw, 1879. Present on nearly all the streams and lakes that furnish fish; extremely abundant at Klamath Lake.

Camp Harney. Bendire.—I obtained a single specimen February 18, 1875. A pair breed on Silvies River and are the only ones I have seen here.

Ridgway. Met with only in the neighborhood of Pyramid Lake where it was rare.

Newberry, 1854. Not rare in California along the San Joaquin and Sacramento Rivers; is very common at the Cascades of the Columbia and at the falls of the Willamette, and still more abundant about the chain of lakes which cover so large a surface in the Klamath Basin. We found it in the Cascade range about the mountain lakes, and indeed in all places where fish, its favorite food, is obtainable.

Heermann, 1854. We found this species in the Tulare Valley on the borders of large lakes, and in one place counted three nests within sight of each other.

Cooper, 1870. Very abundant where not exterminated by the foolish ambition which inspires most gunners to kill an eagle.

Henshaw, 1876. An abundant resident of California particularly along the sea coast. It is also not uncommon in the mountain districts. The islands of the Santa Barbara Channel are the resort of many pairs that remain during the year.

San Diego. T. C. Parker.—January, 1875, a specimen.

L. B.—It is decidedly rare in San Diego county and has not been seen by me in any part of Lower California. It was common in Central California in winter as late as 1866, but is now rare.

Marysville. W. F. Peacock.—November 15, 1885. During the last two years four of these eagles have come into my possession; two of them were shot at Red Bluff about the middle of April, and one was caught in a steel trap at Bangor, Butte county. The fourth was killed in Marysville Buttes and brought to me in flesh, October 20, 1885. I am credibly informed that seven of these birds are seen almost daily in these mountains.

36. Falco mexicanus Schleg. PRAIRIE FALCON.

Camp Harney. Bendire.—Not at all rare during the migration of the water fowl. A few breed in this vicinity.

Henshaw, 1879. A widely diffused species and common in certain localities of this region, as near Camp Bidwell.

Ridgway. A rather common species throughout the Great Basin. It was common during summer about the cliffs of the Ruby Range where the families of young accompanied by their parents flew among the precipitous rocks where they had been bred.

Heermann. I obtained near Sacramento three specimens and saw a young unfledged one in San Francisco. I also procured one on the Farallon Islands.

Alameda and Contra Costa counties. W. E. Bryant.—Rare resident; eggs taken.

Central California. L. B.—Rare winter visitant; probably breeds in Hope Valley, Alpine County, California, on the east slope. San Diego County, winter, rare; mountains south of Campo, May, rare; first seen at Summit, Central Pacific Railroad, September 2, 1885.

37. Falco peregrinus anatum (Bonap.) DUCK HAWK.

San Diego. L. B.—Winter, rare.

Cooper, 1870. I have found this species along the whole southern coast of California, where it resides constantly.

Henshaw, 1876. Appears to be rather common in southern California, being perhaps most so on the coast. It is numerous on the Santa Barbara Islands, also present around Kern Lake.

Alameda and Contra Costa counties. W. E. Bryant.—Rare resident.

Central California. L. B.—Rare winter visitant, but perhaps breeds on Butte Slough, Butte County, where I have seen two or more in June.

Cooper, 1860. Of the western duck hawk I have seen only two pairs, which, in March, 1854, frequented a high wooded cliff at Shoalwater Bay.

British Columbia. John Fannin.—Rare summer resident.

Bendire. Only seen near Malheur Lake attracted by the great number of water fowls of all kinds.

Ridgway. Observed only at Pyramid Lake and along the lower portions of the Truckee River; at the former locality only a single pair.

[I have taken the liberty of placing Dr. Cooper's notes on the duck hawk under this species although they were originally under the head of *Falco nigriceps* Cassin.

Part of the notes included here may belong under the next species but without specimens it is impossible to determine.]

38. **Falco peregrinus pealei** Ridgw.   PEALE'S FALCON.
Habitat: Oregon and northward.

39. **Falco columbarius** Linn.   PIGEON HAWK.
British Columbia.   John Fannin.—Summer resident; not common.

Cooper, 1860. Seems to be rather uncommon in the Territory. I shot one in June, 1853, and did not see another until April, 1855, when they had just arrived at the Strait of Juan de Fuca. It doubtless breeds in the Territory.

Suckley, 1860. About the 1st of August this bird becomes very abundant in the vicinity of Fort Steilacoom, Washington. During the summer and autumn of 1856 I obtained a number of specimens in different stages of plumage.

Walla Walla. Brewster (Bull. N. O. C., vii, 227.)— Varieties *suckleyi* and *richardsonii* collected by Capt. Bendire.

Newberry. We found it paired and nesting about the Klamath Lakes, and it likewise occupies all the region south of the Columbia in Oregon.

Henshaw, 1879. "*Richardsonii*" noted in several localities in Oregon and northern California, and two specimens in immature plumage were taken.

Camp Harney. Bendire.—Rather rare; nest containing young, May, 1876; the only nest I have seen.

Ridgway.   Seen on but three or four occasions.

Central California. L. B.—Rare winter visitant in valleys and foothills; probably very rare summer resident of the fir forests; one seen on Castle Peak near Summit, July 1st, 1885.

Henshaw, 1876. (Var. *richardsonii*). It is found in southern California, and I think not uncommonly, though I took but a single specimen (August 28, at Walkers Basin).

Cooper, 1870. I shot a fine specimen at Fort Mojave, in winter

San Bernardino. F. Stephens.—Rare resident of valley and foothills.

Agua Caliente, San Diego County, Cal. F. Stephens.—One seen April 6 and another April 8, 1886.

San Diego. L. B.—Winter, rare.

40. Falco columbarius suckleyi Ridgw. BLACK MERLIN.

41. Falco richardsonii Ridgw. RICHARDSON'S MERLIN.

[The notes on the Pigeon Hawk were arranged prior to the receipt of the A. O. U. Check List, and at the time I did not know how the different forms would be treated. Having thus far recognized but one form in the few specimens I have collected, I simply follow what I believe to be competent authority.]

42. Falco sparverius Linn. AMERICAN SPARROW HAWK.

San Diego and north Lower California; tolerably common resident.—L. B.

Volcan Mountains. W. O. Emerson.—March 9, first noticed; Santa Isabel, April 3, in pairs.

San Bernardino. F. Stephens.—Tolerably common resident in the valley and foothills.

Agua Caliente. F. Stephens.—Several seen; probably resident in small numbers.

Henshaw, 1876. In California, as throughout the west generally, very numerous. Great numbers of these birds near Santa Barbara in July.

Cooper, 1870. Resides constantly in California, frequenting chiefly the plains.

Alameda and Contra Costa counties. W. E. Bryant.—Tolerably common resident.

Central California. L. B.—Common resident of the valleys and foothills; common about subalpine meadows in the Sierra in summer.

Heermann. Abundant throughout the whole of California.

Newberry. In the Sacramento Valley, in the interior basin, and in the mountains and valleys of Oregon, we found it everywhere quite as abundant as in the eastern states.

Willamette Valley. O. B. Johnson.—Very common; nesting in holes, usually of a woodpecker.

Cooper, 1860. Extremely common during summer about prairies, even at the summit of the Cascade Range, but I have never observed it in the forests or near the sea shore. I noticed their arrival at Puget Sound early in May.

Suckley, 1860. Exceedingly abundant on the Nisqually Plains, Puget Sound.

British Columbia. John Fannin.—Abundant summer resident, arriving at Burrard Inlet, April 13, 1855. The species was again seen April 17, and by May 15 it was common. It is common here in breeding season.

Henshaw, 1879. Very common throughout this whole region.

Bendire. A very common summer resident, breeding abundantly about Camp Harney.

Ridgway. Regarding the western range of this widely distributed species, nothing more need be said than that it occurs everywhere in suitable places.

Hoffman. Generally abundant in the more favorable localities of upper Nevada.

43. Polyborus cheriway (Jacq.) AUDUBON'S CARACARA.

Heermann. I am happy to be able to add this interesting species to the fauna of California, having seen it on the Colorado River near Fort Yuma.

44. Pandion haliaëtus carolinensis (Gmel.) AMERICAN OSPREY.

A few seen at San Diego in winter.—L. B.

Santa Catalina Island. F. Stephens.—August, 1886, one seen.

Cooper, 1870. Found wherever there is clear water containing fish. June 20 the young birds about Catalina Island appeared to be fully fledged.

Santa Cruz. Joseph Skirm.—A pair have nested here several summers.

Bodega Bay. F. H. Holmes.—Shot here.

Newberry. Along the Sacramento River, far up on Pit River, around the Klamath Lakes, in the Cascade Mountains, on the Columbia and Willamette, we still found the fish-hawk.

Heermann. Abundant, being met with throughout the whole extent of California. In the fall it migrates south.

O. B. Johnson. Common along the Columbia and Willamette Rivers, nesting in trees.

Cooper, 1860. Common along the coast, arriving at Puget Sound by the middle of April.

British Columbia. John Fannin.—Abundant summer resident.

Henshaw, 1879. Present on nearly all the lakes and streams that furnish fish.

Camp Harney. Bendire.—A rare summer visitor.

Ridgway. Seen only along the lower portion of the Truckee River near Pyramid Lake where it was rather common in May. It no doubt bred in that locality.

45. Strix pratincola (Bonap.) AMERICAN BARN OWL.

San Diego. L. B.—Seen once, but not seen in any part of Lower California.

Poway. F. E. Blaisdell.—Common resident.

Cooper, 1870. Abundant throughout the southern part of California, especially near the coast.

San Bernardino. F. Stephens.—Common resident of the valley.

Agua Caliente, San Diego County, Cal. March 18 to April 15, 1886. Heard nearly every night; probably a resident pair.

Henshaw, 1876. Appears to be common throughout southern California and in some portions, as in the swamps near Los Angeles, and again in the San Bernardino Valley, I found it in great numbers.

Santa Cruz. Joseph Skirm.—Common along the coast, mostly nesting in holes in the cliffs.

San Jose. A. L. Parkhurst.—Fresh eggs January 25, 1885, and young found in a nest February 8.

Alameda and Contra Costa counties. W. E. Bryant.—Common resident.

Central California. L. B.—Rather common in the valleys, especially in willows along the rivers.

Newberry. Apparently more abundant on the western coast than in the Atlantic States, and more common in California than in Oregon. It also inhabits the Klamath Basin.

Henshaw, 1879. At Camp Bidwell it seemed to be a tolerably common species. That it occurs on the east side of the Sierras I obtained ample proof during the past season.

46. Asio wilsonianus (Less.) AMERICAN LONG-EARED OWL.

San Diego and northern 100 miles of Lower California, tolerably common resident.—L. B.

Poway. F. E. Blaisdell.—Resident.

San Bernardino. F. Stephens.—Rare resident of the valley.

Los Angeles. Henshaw, 1876.—Seen several times.

Santa Paula, Ventura County. B. W. Evermann.—February 16, 1881, laying already.

Alameda and Contra Costa counties. W. E. Bryant.—Rare resident.

Sebastopol. F. H. Holmes.—Collected here.

Central California. L. B.—Apparently rare; seen only on a few occasions in the valleys and once in the foothills. I have several times seen three or four in the high Sierra, perhaps migrants, but possibly summer residents of these mountains.

Ridgway. Seldom if ever did we enter a willow copse of any extent without starting one or more. This was the case both near the Sacramento and in the interior, and in summer as well as in winter.

Cooper, 1860. Obtained only once on the banks of the Columbia near the Dalles, November, 1853.

British Columbia. John Fannin.—Summer resident; not common.

Henshaw, 1879. Numerous in the thickets of the low lands, where it is resident throughout the year.

Camp Harney. Bendire.—Moderately abundant and resident, frequenting the dense thickets along streams, and here constructing their own nests.

Hoffman. Very common in every favorable locality. Near Carlin and at various localities west and south of that place.

Cooper, 1870. Wanders into the barren, treeless deserts east of the Sierra Nevada.

47. **Asio accipitrinus** (Pall.) SHORT-EARED OWL.

F. E. Blaisdell.—One shot at Temecula, November 13, 1883; I have not seen it at Poway.

San Bernardino. F. Stephens.—Rare accidental visitant to the valley.

Cooper, 1870. I have not seen them south of Santa Clara Valley.

Vicinity of Oakland. W. E. Bryant.—Tolerably common winter visitant.

Central California. L. B.—Common in winter in tule marshes; was last seen at Stockton about April 1, 1878, and reappeared on September 30, following.

Cooper, 1860. In fall and winter it appears in large numbers on the low prairies of the coast. I have not observed it during summer in the territory.

British Columbia. John Fannin.—Summer resident; not common.

Henshaw, 1879. Common in the sedgy marshes about Warner Lake, Oregon.

Bendire. Only a summer resident, confining itself to the marshes near Malheur Lake. Two sets of eggs May 28, on the ground in a large swamp.

Heermann. I started from the bushes a specimen on the desert extending between the Tejon Pass and the Mojave River, on the borders of which I also met another.

Newberry. Found throughout California and Oregon; especially common in the Klamath Basin. Upper Pit River, found in considerable numbers.

48. Syrnium occidentale Xantus. SPOTTED OWL.

Cooper, 1870. It was discovered by Mr. J. Xantus at Fort Tejon, March 6, 1858, but only one specimen was obtained.

L. B.—Common at Big Trees, Calaveras County, and vicinity in summer, and perhaps in winter. It frequents the densest parts of the fir forests. June 13, 1882, male and female shot together in early evening; stomachs empty; ovaries small. Both appeared to be adults, and were beginning or concluding moult.

49. Scotiaptex cinerea (Gmel.) GREAT GRAY OWL.

Newberry. We obtained proofs of its existence in the Sacramento Valley, in the Cascade Mountains, in the Des Chutes Basin, and on the Columbia.

Mr. Wm. Proud has a specimen, which he informed me was brought to him in flesh soon after it had been shot, in the hills near Chico.—L. B.

Willamette Valley. O. B. Johnson. — Occasionally seen in heavily wooded districts.

Cooper, 1882. Common in the dense spruce forests near the Columbia River and northward.

Baird, Brewer and Ridgway. A single specimen of this owl was taken at Sitka by Bischoff; and on the 20th of April Mr. Dall obtained a female that had been shot near Mulato. He subsequently obtained several specimens in that region.

50. Nyctala acadica (Gmel.) SAW-WHET OWL.

British Columbia. John Fannin.—Common resident.

Vancouver. Cooper, 1860. February 3, 1854, I found one dead.

Suckley, 1860. A fine specimen found at the Dalles of the Columbia in December, 1853.

Willamette Valley. O. B. Johnson.—I have a single example.

Camp Harney. Bendire.—Only taken during very cold weather in the winter. I think it a resident, and that it breeds here.

Ridgway. But a single specimen was met with—at Thousand Spring Valley, Nevada, September 24.

Dr. Gambel met with it in California (northward?).

Fort Tejon. Baird, Brewer and Ridgway. Female specimen collected by J. Xantus.

Cooper, 1870. I have seen in the German Academy of Natural Sciences a specimen of *Nyctale albifrons* brought from Nevada close to the boundary of California.

Lately *N. albifrons* has become known as the young of *N. acadica*. At Lake Tahoe, September 21, 1889, a young male was taken by myself.—L. B.

51. Megascops asio bendirei (Brewst.) CALIFORNIA SCREECH OWL.

Cooper, 1870. Quite common in the wooded parts of the State.

Poway. F. E. Blaisdell.—Occasionally seen.

San Bernardino. F. Stephens.—Tolerably common resident of the valley and foothills.

Henshaw, 1876. A common resident of California.

Santa Cruz. Joseph Skirm.—Quite common.

Alameda and Contra Costa counties. W. E. Bryant.— Tolerably common resident.

Central California. L. B.—Common resident in the valley; less common in the foothills; not seen or heard in the fir forest.

Willamette Valley. O. B. Johnson.—Very common, breeding in trees.

[Perhaps the Willamette birds are *M. a. kennicottii*.]

British Columbia. John Fannin.—Common resident. [*M. a. kennicottii?*]

Suckley, 1860. A specimen in the mottled plumage was obtained by me at Fort Vancouver. [*M. a. kennicottii?*]

52. Megascops asio kennicottii (Elliott). KENNICOTT'S SCREECH OWL.

Beaverton, Washington County, Oregon. A. W. Anthony (Auk, April, 1886.)—A not uncommon resident.

Brewster. (Bull. Nutt. Orn. Club, vii, 27). Specimens from Fort Walla Walla, W. T., from John Day River and Portland, Oregon; Sitka, Alaska; Idaho and Montana.

53. Megascops flammeolus (Kaup.) FLAMMULATED SCREECH OWL.

Baird, Brewer and Ridgway. A specimen collected at Fort Cook by Captain John Feilner. This was a young bird, evidently raised in that locality.

Big Trees, August 16, 1880. L. B.—One specimen, which I sent to the Smithsonian Institution. It was shot about 10 o'clock P. M. Its stomach contained fragments of beetles. It had one or more companions. It is probably not rare in the denser parts of the forests of the Sierras in summer.

54. Bubo virginianus subarcticus (Hoy.) WESTERN HORNED OWL.

British Columbia. John Fannin.—Common resident.

Cooper, 1860. Common constant resident in all parts of the territory.

Willamette Valley. O. B. Johnson.—Quite common.

Central California. L. B.—Tolerably common summer resident of the forests; rare summer resident of the valleys; more common in the latter in winter.

Alameda and Contra Costa counties. W. E. Byrant.—Tolerably common resident; several pairs breed near Berkeley annually.

Berkeley. T. S. Palmer.—Accidental visitant.

Santa Cruz. Joseph Skirm.—Rare.

Fort Tejon. Henshaw, 1876.—August 8.

San Bernardino Mountains. F. Stephens.—Rare resident.

Volcan Mountains. W. O. Emerson.—An individual seen January 29, and again in February.

Poway. F. E. Blaisdell.—Tolerably common resident.

[Mr. B. W. Evermann, of Santa Paula, informed me February 16, 1881, that the species was then breeding there.]

Henshaw, 1879. Often heard in the mountains where it is resident. Two individuals of the many seen were obtained in the Cascade Mountains. They represent the common interior type.

Camp Harney. Bendire.—Several specimens taken at different seasons are referable to var. *arcticus*, while differing in coloration.

Brewster. (Bull. Nutt. Orn. Club, October, 1882.)—At Fort Walla Walla, during the autumn of 1881, Captain Bendire secured no less than fourteen specimens of which twelve are now before me. In a general way these are referable as follows: Eight to *saturatus;* two to *subarcticus*, and two to a form apparently about intermediate between these races.

Ridgway. Found by us in all wooded districts excepting the Sacramento valley. Near Pyramid Lake it was abundant in December. It was also common near Carson. Nest and three eggs (of *subarcticus*), April 21.

55. **Bubo virginianus saturatus** Ridgw. DUSKY HORNED OWL.

56. **Nyctea nyctea** (Linn.) SNOWY OWL.

British Columbia. John Fannin.—Rare winter resident.

Brewster. (Bull. Nutt. Orn. Club, vol. vii.)—Walla Walla; collected by Captain Bendire.

Willamette Valley. O. B. Johnson.—Occasionally killed in winter by hunters.

Camp Harney. Bendire.—Rare winter visitor.

57. Surnia ulula caparoch (Müll.) AMERICAN HAWK OWL.

British Columbia. John Fannin.—Rare summer resident.

58. Speotyto cunicularia hypogæa (Bonap.) BURROWING OWL.

Common in many parts of San Diego County.—L. B.

Poway. F. E. Blaisdell.—Common resident.

San Bernardino. F. Stephens.—Tolerably common resident of the valley.

Henshaw, 1876. Nowhere in the west does this owl occur oftener, or in greater numbers, than in southern California.

Santa Cruz. Joseph Skirm.—Common.

Alameda and Contra Costa counties. W. E. Bryant.—Rare resident; formerly common.

Cooper, 1870. Probably one of the most common birds in California.

Central California. L. B.—Very common resident in the valleys and lower foothills as it is, no doubt, in those throughout the State.

Klamath Basin. Newberry.—Less frequent in the Sacramento Valley.

Suckley, 1860. Abundant at the Dalles; not seen by me west of the Cascade Mountains.

British Columbia.—John Fannin.—Rare summer resident east of the Cascades.

Henshaw, 1879. (East slope). Numerous in all suitable localities throughout this region.

Camp Harney. Bendire.—A common summer resident, arriving about the end of March.

Hoffman. Near Antelope Creek, about 60 miles north of Battle Mountain, two individuals were seen.

Ridgway. It was most numerous on the dry plains

near Sacramento. Eastward of the Sierra Nevada we found it only at wide intervals. It was rare about Carson. A single pair was seen on the mesa between the Humboldt River and the west Humboldt Mountains, and a few were noticed in Fairview Valley, while in the neighborhood of Salt Lake City it was more common.

Cooper, 1870. I found one living near the Colorado, in a burrow which it had apparently just made.

59. **Glaucidium gnoma californicum** (Scl.) CALIFORNIA PIGMY OWL.

British Columbia. John Fannin.—Common resident.

Cooper, 1860. Only once seen. On the 1st of November, 1854, I observed it among a flock of sparrows, which did not seem at all frightened by its presence.

Willamette Valley. O. B. Johnson.—Quite common.

Newberry. I procured specimens on the Cascade Mountains in Oregon.

Camp Harney. Bendire.—Moderately abundant in winter, and some unquestionably breed.

Calaveras Big Trees. L. B.—July 4, 1882, juveniles just out of the nest. It is quite common in the foothills of Central California in winter, and is a rare summer resident of the fir forests. Mr. Charles A. Allen, of Nicasio, informed me that he had taken twenty-five specimens there in one season, if I remember correctly. Mr. William Proud shot one near Chico, February 7, 1885, nearer sea level than I had previously seen it in the interior of the State.

Santa Cruz. Joseph Skirm.—May, 1881, I found a nest with three eggs. Mr. Ready also found a nest at Santa Cruz.

60. **Micropallas whitneyi** (Cooper). ELF OWL.

Cooper, 1870. No. 208, State Collection; shot at Fort Mojave, Colorado Valley, April 26, 1861, is, as yet, a unique specimen.

Captain Bendire secured several specimens near Tucson. Mr. F. Stephens found it a very common bird about Tucson and Camp Lowell (Brewster Bull. Nutt. Orn. Clüb, vol. viii, 27), also stating that Mr. Stephens found several nests in deserted woodpecker's holes in the giant cactus.

61. **Geococcyx californianus** (Less.) ROAD-RUNNER.

Common at San Diego; first notes here, January 19, 1884.—L. B.

Poway. F. E. Blaisdell.—Common resident.

San Bernardino. F. Stephens.—Rare resident of the valley and foothills.

Henshaw, 1876. Abundant resident through Southern California.

Cooper, 1870. At Santa Barbara I observed a young one nearly fledged in May.

Contra Costa County. W. E. Bryant.—Rare resident.

Berkeley. T. S. Palmer.—Rare; accidental visitant.

San Rafael. A. M. Ingersoll.—Very rare.

Central California. L. B.—Rare resident; seldom found above fifteen hundred feet.

Murphys. John J. Snyder.—September, 1885, a specimen.

Sebastopol. F. H. Holmes.

Igo, Shasta County. E. L. Ballou.—Rare resident.

Hoffman. I found this bird only in the Colorado Valley, in the vicinity of Fort Mojave, where it was not of uncommon frequency.

62. Coccyzus americanus occidentalis Ridgw. CALIFORNIA CUCKOO.

Poway. F. E. Blaisdell.—Once seen in 1875 and once in 1876.

San Jose. A. L. Parkhurst.—June 6, 1885, once seen and heard.

Oakland. W. E. Bryant.—One specimen taken here by Charles N. Comstock.

Sebastopol. F. H. Holmes.—Shot here August 8, 1884; seen June 13, 1885. Very rare.

Cooper, 1870. While stationed at Sacramento, in 1865, I found these birds quite common in the large cottonwood trees about the city from about May 1 to September 1. In the summer of 1862 Mr. Gruber obtained two specimens, shot in Napa Valley.

Ridgway—At Sacramento City its well-known notes were heard on more than one occasion in June, among the oak groves in the outskirts of the city, while across the Sierra Nevada, several individuals were seen in July in the wooded valley of the Lower Truckee.

L. B.—Marysville, June, 1878, common in the willow and poplar thickets. July 7, 1884, two juveniles just out of nest.

Marysville. W. F. Peacock.—May 19, 1884, first; June 6 next seen; tolerably common; breeds. First seen May 1, 1885; common May 8.

Chico. Wm. Proud.—June 4, 1884, first seen; usually very rare here.

Willamette Valley. O. B. Johnson.—Rare; two specimens killed in this vicinity.

Bendire. Keeny's Ferry, Oregon side of Snake River, August 2, 1876, a nest containing half grown young birds.

Kamloops, B. C. John Fannin.—*C. americanus* found nesting.

63. Ceryle alcyon (Linn.) BELTED KINGFISHER.

San Diego. Resident, not numerous; never abundant.—L. B.

Poway. F. E. Blaisdell.—Common along the coast.

San Bernardino. F. Stephens.—Very rare summer resident of the valley.

Agua Caliente, San Diego County, Cal. F. Stephens. Seen almost every day from April 3 to April 15, 1886 (day of leaving.)

Henshaw, 1876. Every small stream which is stocked with fish is occupied by one or more of these birds.

Contra Costa County. W. E. Bryant.

Central California. L. B.—Tolerably common on clear streams which contain fish. One observed catching trout, near Big Trees, January, 1879, snow two feet deep. Summit, autumn, 1885, several seen, apparently migrants, perhaps going from the east slope to California to winter.

Cooper, 1870. Common along the coast.

Willamette Valley. O. B. Johnson.—Constant resident.

Cooper, 1860. Abundant throughout the year along every stream as well as the coast; probably does not retire southward, except in those uncommon winters when all the fresh water becomes frozen.

British Columbia. John Fannin.—Very abundant resident.

Henshaw, 1879. Of common occurrence on all the fish-stocked streams.

Camp Harney. Bendire.—A rare summer visitor.

Ridgway. Found in the vicinity of all streams and lakes containing fish. In the lower valleys it was resident, but in the mountains it was found only in summer.

Hoffman. Common in favorable localities.

64. Ceryle cabanisi (Tschudi.) TEXAN KINGFISHER.

Baird, Brewer and Ridgway. Dr. Coues states that they have been observed at several points on the Colorado River, between Fort Mojave and Fort Yuma—the only instances of their occurrence in the United States other than on the Rio Grande.

65. Dryobates villosus harrisii (Aud.) HARRIS' WOODPECKER.

Volcan Mountains. W. O. Emerson.—March 24 one male; the only individual seen.

Volcan Mountains. F. E. Blaisdell. — August 21; November 28 rather common.

San Bernardino. F. Stephens.—Breeds in the pine region of the mountains; rare winter visitant to the foothills and valley.

Henshaw, 1876. More or less common summer resident of the mountains. Specimens from Fort Tejon and near Mount Whitney.

Cooper, 1870. Resident as far south as Santa Barbara. I found them more common in the higher coast range near Santa Cruz.

Contra Costa County. W. E. Bryant.—One taken.

Central California. L. B.—Tolerably common summer resident of the fir forests, apparently driven down from the higher Sierra in winter, though I saw one at Donner Lake, November 16, 1884, and a few at Big Trees, January, 1879, when there was but little snow. It is rarely met in the valleys in winter. Its burrows are sometimes within three feet of the ground; eggs, usually four, though I have found seven, one apparently unfruitful.

Willamette Valley. O. B. Johnson.—Common resident; breeding extensively.

Cooper, 1860. The most abundant species in the Territory; a constant resident.

Suckley, 1860. A winter resident of Fort Dalles and Fort Steilacoom.

Henshaw, 1879. Numerous as a resident of the pineries.

Camp Harney. Bendire. — Sparingly distributed through the forests of the Blue Mountains in summer; in spring and fall they frequent the shrubbery along creeks in the valleys and a few winter in such localities.

Ridgway. Met with in all wooded localities throughout the year, from the Sierra Nevada eastward.

Hoffman. Not uncommon in the timbered districts over the greater portion of Nevada; none were seen in the southern regions.

66. Dryobates pubescens (Linn.) DOWNY WOODPECKER.

Marysville, winter of 1877-78, a typical specimen; same locality, December 21, 1884, typical female; on the following day another typical female and one nearly so. These latter were the first and only ones of either the downy or Gairdner's woodpeckers shot here in 1884, and they were not selected. Mr. A. W. Anthony sent me one he shot at Beaverton, Oregon, and that, too, was nearly typical. The specimen of 1877-78 is No. 73,606 of the Smithsonian collection.—L. B.

67. Dryobates pubescens gairdnerii (Aud.) GAIRDNER'S WOODPECKER.

San Bernardino. F. Stephens.—Rare accidental visitant to the valley.

Henshaw, 1876. Not so arboreal as the Harris's woodpecker. We do not find it among the high mountains of California, save occasionally, but with the Nuttall's it resorts to the low districts and frequents to a great extent the deciduous timber, especially the oaks. Santa Barbara, July 6, juvenile.

Cooper, 1870. Santa Clara Valley, May 24, 1864, nest containing young.

Alameda and Contra Costa counties. W. E. Bryant.—Rare resident.

Berkeley. T. S. Palmer.—Rare visitant. December 22, 1885, one female, again seen on April 14 and May 31, 1885.

Central California. L. B.—Rare in the fir forests at all times; more common in foothills and valleys; common in breeding season in willow thickets of the Sacramento Valley at several points, perhaps generally so.

Newberry. Very common in northern California and Oregon.

Willamette Valley. O. B. Johnson.—Abundant; nesting in tops of dead willows.

Cooper, 1860. Always found among the oaks, maple and alders.

Suckley, 1860. Extremely common on the lower Columbia, especially among the willow trees.

British Columbia. John Fannin.—Common resident.

Henshaw, 1879. Along the eastern slope as everywhere throughout the middle region, this is a rare species and but a single individual was seen.

Camp Harney. Bendire.—Only met in the John Day River Valley, Oregon, and it appears to be rare there.

Ridgway. Unaccountably rare in all portions of the country, even where its larger cousin, *D. harrisii*, abounded; indeed it was seen at only two localities along the entire route (specimens taken at upper Humboldt Valley).

68. Dryobates scalaris bairdi (Scl.) BAIRD'S WOODPECKER.

Heermann. Abounding in the woods about Fort Yuma.

Cooper, 1870. Abundant in the Colorado Valley [Fort Mojave?].

[I believe there are no records of its occurrence in California west of the Sierra Nevada, nor of its occurrence in any part of Lower California, though it is likely to be found in the northeastern part of the latter. By the Sierra Nevada I wish to signify the mountain chain which extends at least as far south as lat. 31° in Lower California and which separates the Colorado Desert from the coast region in California, includes the Cuayamaca, Volcan, San Jacinto, San Bernardino Mountains, and connects with or continues as the Cascade Mountains. A striking feature of these mountains along the head of the Gulf of California is that they are very rocky and steep].

69. **Dryobates nuttallii** (Gamb.) NUTTALL'S WOODPECKER.

Rather rare resident about San Diego.

Tehachapi. April, 1889, rather common.—L. B.

Volcan Mountains. F. E. Blaisdell.—August 21, November 28, 1884, frequently noticed.

Santa Isabel. W. O. Emerson.—Seen from January 23, 1884, to April 3.

San Bernardino. F. Stephens.—Rare resident of the valleys and foothills; found nesting in the valley.

Henshaw, 1876. Ranges from the lower valleys up into the mountains to a height of at least 6,000 feet, where, near Fort Tejon, I found it fairly numerous among the pines, this being the only locality where I found it among the conifers.

Cooper, 1870. It frequents the oaks and smaller trees almost exclusively, avoiding the coniferous forests. I have not observed it west of the coast range except at Santa Barbara, nor have I seen any around gardens or orchards.

Alameda and Contra Costa counties. W. E. Bryant.—Tolerably common resident.

Central California. L. B.—Common resident of the valleys and foothills. I have never seen it above 4,000 feet, and rarely above 3,000 feet, and then not in breeding season. It is resident and tolerably common as far north as Red Bluff.

70. **Xenopicus albolarvatus** (Cass.) WHITE-HEADED WOODPECKER.

Volcan Mountains. F. E. Blaisdell.—August 21, 1884; rare. [The most southern California record.]

San Bernardino. F. Stevens.—Breeds on the high mountains.

Henshaw. Tolerably numerous in the pine woods of the mountains near Fort Tejon, and also in the Mt. Whitney region.

Central California. L. B.—Common in the fir forests of the Sierra from about 4,000 feet upward; most numerous at about 5,000 feet. Rare in the tamaracks *(Pinus contorta)* at Blood's, Summit, Sierra City and Butte Creek House; a few seen at Sierra Valley, Donner Lake, Lake Tahoe, and other localities on the east slope. Its burrows are often within two or three feet of the ground. I have seen two nests in cuts for shakes or shingles, made after the tree was sawed into sections, and one in a small, short stub of dogwood *(Cornus nuttalli)*. May 25, 1879, first full set of eggs taken at Big Trees; June 6, 1880, first set. At Blood's, 7,200 feet altitude, I have taken them as late as July 17. The eggs are usually four, although I have seen five. In winter it is found sparingly in the upper edge of the foothills at about 3,000 feet altitude. I found it rather common about Big Trees in the mild January of 1879 until two feet of snow fell, after which none were seen.

Mr. John Snyder, of Murphys, says he found the first one there in the fall of 1885 on October 18th.

Newberry. We found this species only in the Cascade Mountains of Oregon, where it is, apparently, not common.

Cooper, 1870. Common at Dalles, Columbia River. I have also found it as far north as Fort Colville, Washington, near lat. 49°.

Henshaw, 1879. Along the eastern slope it occurs here and there as a resident.

Camp Harney. Bendire.—Not common; only met with in the pine forests of Blue Mountains; remains here throughout the year.

British Columbia. John Fannin.—Rare summer resident.

Ridgway. It was common near Carson throughout the winter, keeping entirely among the pines, though sometimes coming down to the lower edge of the woods.

71. Picoides arcticus (Swains.) ARCTIC THREE-TOED WOODPECKER.

Blood's, on Big Trees and Carson road, rare summer resident; a pair feeding young here June 20, 1881, in a dead tree about eight feet from the ground. Summit, rare summer resident. Butte County, lat 40° 10', or thereabout, altitude 6,700 feet, two pairs July 1–3, 1884.—L. B.

Ridgway. The only specimen seen was the one obtained at Carson, February 19, 1868.

Newberry. This woodpecker we found only in the Cascade Mountains, within a hundred miles of the Columbia.

Henshaw, 1879. A rather common and constant resident of the pine woods from Carson northward into Oregon.

Near Soda Springs, Blue Mountains. Bendire.—I have seen what I take to be this species on two occasions. (Proc. Bost. Soc. Nat. Hist., March 21, 1877.)

British Columbia. John Fannin.—Rare resident.

72. Picoides americanus dorsalis Baird. ALPINE THREE-TOED WOODPECKER.

British Columbia. John Fannin.—Summer resident; not common.

Alaska. Fort Yukon. Lucien M. Turner.—Numerous.

Southern Alaska. E. W. Nelson.

73. Sphyrapicus varius nuchalis Baird. RED-NAPED SAPSUCKER.

January 23, 1884, a male specimen taken 35 miles east of San Diego, on the west slope of the mountains; others probably seen.—L. B.

San Bernardino Mountains. F. Stephens. — Rare winter visitant.

Henshaw. This middle region form extends across from the Rocky Mountains, and occurs in summer along the eastern slope. Mr. Ridgway alludes to the fact that in the region between the Rocky Mountains and the Cascade Range specimens of this form give evidence, by the admixture of red in the black auricular stripe, the black pectoral collar, and in the white area surrounding it, of the change soon to result in the variety *ruber*—however, two males taken in the Warner Mountains, northern California, are not distinguishable from Rocky Mountain specimens.

British Columbia. John Fannin.—Common resident.

Camp Harney. Bendire.—Moderately common in the Blue Mountains where it breeds; not a winter resident.

Hoffman. Rather common in suitable localities throughout the northern regions. (Nevada).

Ridgway. Between the Sierra Nevada and Rocky Mountains; common in suitable localities. One was obtained on the eastern slope of the Sierra Nevada, near Carson, April 4, 1868. Very rare throughout western Nevada, but became abundant in the eastern portion of the State.

Cooper, 1870. Fort Mojave, February 20, 1861, a female specimen—the only one I saw.

Fort Yuma. Heermann.—Not rare.

74. Sphyrapicus ruber (Gmel.) RED-BREASTED SAPSUCKER.

Big Laguna, San Diego County. F. E Blaisdell.— November 12, a specimen.

Henshaw, 1876. It breeds about as far south as Fort Tejon, as I took a young bird in the mountains in August, and saw several more. Later, in October, I took a pair near Kernville.

Cooper, 1870. I have not seen any south of Santa Clara, and there only in the mountains of the Coast Range in early spring.

Contra Costa and Alameda counties. W. E. Bryant.— Rare winter visitant.

Berkeley. T. S. Palmer.—Tolerably common accidental visitant. Two or three individuals seen every winter. January 30, February 1 and March 10, 1886.

Sebastopol. F. H. Holmes.—Tolerably common winter visitant. October 9, 1884, first; March 17 following, last seen.

Central California. L. B.—Common summer resident in the fir forest of the Sierra. Like nearly all the Californian woodpeckers it is found lower down in winter, then becoming rather common in the foothills, although rarely seen in the valleys. I have seen a considerable number of its burrows in Calaveras, Tuolumne, Alpine, Butte, and other counties. They are rarely below 30 feet, and are often overlooked or found with some difficulty, as they frequently are in bark-covered trees. It was noticed at Sierra Valley and Donner Lake.

Willamette Valley. O. B. Johnson.—Not very common. I found a nest in a cottonwood stub about 30 feet from the ground.

Cooper, 1860. I have only met with them three times, in spring and fall.

Suckley, 1860. I have seen but one specimen.

British Columbia. John Fannin.—Transient visitant; not common.

Ridgway. Observed only in the Sierra Nevada, chiefly on the western slope of that range.

Henshaw, 1879. I was able to obtain no evidence that this variety summers along the eastern slope, and am compelled to believe that it is only found here in the character of a fall and winter visitor.

75. **Sphyrapicus thyroideus** (Cass.) WILLIAMSON'S SAPSUCKER.

Henshaw, 1876. Quite common in the heavy pine and redwood forests near Mt. Whitney in September, and they doubtless breed there.

Central California. L. B.—Tolerably common from about 7,000 feet upward in summer often breeding in living tamaracks and covered with their resinous juice. In winter, down to about the lower edge of the sugar pines *(Pinus lambertiana)*, altitude about 2,500 feet, but rare here and mostly female or young birds found so low, while at Big Trees, January 6–13, 1879, I got thirteen adult males. In breeding season they are most numerous in the tamaracks in valleys as at Blood's, Hermit Valley, Blue Lakes, etc. Their burrows vary from five or six feet up to thirty or forty feet. The young were still in the burrows at Blood's, July 21, 1880, but in 1881 they were about a month earlier. The eggs, as usual with woodpeckers, are glossy white. In Ornithology of the Geological Survey of California the male is called "Williamson's woodpecker," the female the "round-headed woodpecker." Mr. Henshaw first discovered that they were the same species.

Ridgway. Both on the Sierra Nevada and in the Wasatch. It was a winter resident among the pines near Carson.

Henshaw, 1879. Of rather frequent occurrence all through the mountains.

Hoffman. Found throughout the elevated coniferous regions.

Cooper, 1870. The first specimen that I met with was a straggler in winter to the Colorado Valley.

76. **Ceophlœus pileatus** (Linn.) PILEATED WOODPECKER.

Henshaw, 1876. Found in the Sierra as far south as latitude 37°, where I saw two individuals in October.

L. B.—Not rare in the densest pine and fir forests of Central California where it is probably resident, as I saw two at Big Trees, January 16, 1879, when snow was two feet deep. Eel River, Mendocino County, common.

Igo, Shasta County. E. L. Ballou.—Common resident.

Willamette Valley. O. B. Johnson.—Common in heavy timbered districts.

Cooper, 1860. An abundant and constant resident in the forests of the territory.

Suckley, 1860. Abundant in the vicinity of Fort Steilacoom in summer.

British Columbia. John Fannin.—Abundant resident.

Henshaw, 1879. Not noted by our party at any point along the eastern slope.

Camp Harney. Bendire.—Seen but twice.

77. Melanerpes formicivorus bairdi Ridgw. CALIFORNIAN WOODPECKER.

Common from Campo to Hansens in May, 1884; not noted on the route to San Pedro Mountain via San Rafael. Common in oaks in several parts of San Diego County. Common resident throughout the interior of San Diego County.—L. B.

Volcan Mountains. F. E. Blaisdell.—August 21 to November 28, 1884, abundant.

Volcan Mountains. W. O. Emerson.—Winter, common but not visible in snow storms when they took refuge in their burrows in the oaks.

San Bernardino Mountains. F. Stephens.—Tolerably common to common resident.

Henshaw. 1876. The habitat of this woodpecker in California, as in Arizona, seems to be determined by the range of the oaks. Fort Tejon in August.

Contra Costa County. W. E. Bryant.—Tolerably common resident.

Berkeley. T. S. Palmer.—Tolerably common winter visitant.

Sebastopol. F. H. Holmes.

Newberry. The range of this species extends to the Columbia and perhaps above, to the westward of the Cascade Range, though more common in California than in Oregon.

Baird, Brewer and Ridgway. Mr. Lord met with it in abundance on his journey from Yreka to the boundary line of British Columbia.

[It is not in Mr. Johnson's list of birds of Willamette Valley; Drs. Cooper and Suckley did not see it in Washington Territory, and Mr. Henshaw did not find it on the east slope where I believe it has never been collected. It is resident in Central California below the firs, occasionally wandering to a height of 5,000 feet in the Sierra after breeding; tolerably common at Red Bluff in winter and probably outnumbers all the other woodpeckers in California. Specimens shot at Gridley, February 23, had eaten, principally, acorns, which the same individuals had probably stored, in their customary manner, in a dead oak near my residence.]

78. **Melanerpes torquatus** (Wils.) LEWIS'S WOODPECKER.

Poway. F. E. Blaisdell.—I saw but one here in 1883. I observed the species at Temecula, November 14, 1883. It was abundant in Volcan Mountains during September and October, 1884.

San Bernardino Mountains. F. Stephens.—Rare resident.

[Mr. Henshaw did not find it in summer about Los Angeles and Santa Barbara, as he says: "I did not see the species until reaching Fort Tejon in August. It was here and in other places, in the pineries, common in certain localities."]

Tehachapi. L. B.—Very common in spring.

Cooper, 1870. Quite common near New Almaden, but not elsewhere in the Coast Range southward during summer.

Alameda and Contra Costa counties. W. E. Bryant. Rare resident.

Sebastopol. F. H. Holmes.—An abundant accidental visitant. This species appeared September 16, 1884, in great numbers, flying high from the north, coming in straggling parties. They soon afterward settled down in the valley and appeared to be at home. They were abundant until about May 1 when they commenced to leave, and by May 10 they were all gone. They were continually storing up acorns in the dead oaks, after the manner of the Californian woodpecker, and were not at all shy.

Ukiah, Mendocino County. George E. Aull.—Every six or seven years we have an inundation of Lewis's woodpeckers. Last autumn they fairly swarmed, and though the most of them left at the beginning of winter, yet there are still a few here (February 20, 1885).

Central California. L. B.—Very common in winter; moderately common in certain localities in summer. I have not found it breeding on the west slope above 3,000 feet, though a summer resident on the east slope at a much greater elevation. Many of these no doubt cross the Sierra and spend the winter in California; and perhaps Mr. Henshaw's Fort Tejon birds were migrants. At Summit, August 16, 1885, I noticed them crossing to the southwest, and afterward on several occasions as late as September 7th, all going in a southwesterly direction.

Newberry.—I first saw it in Lassens Butte, in Northern California. Subsequently we noticed them in the mountains all the way to the Columbia.

O. B. Johnson, 1880. Common along the Columbia in winter, a few remaining to breed.

Cooper, 1860.—Abundant in summer in all the interior districts, never approaching the cooler coast where few of the trees grow which it pefers to inhabit.

British Columbia. John Fannin.—Rare summer resident east of the Cascades.

Henshaw, 1879.—Nowhere in its wide range more abundant than at the base of the eastern slope, through

Nevada, California and Oregon. Seen at the Dalles the last of October.

Camp Harney. Bendire.—A very common summer resident, breeding abundantly; arrives here about May 1, remains until the middle of October. Usual number of eggs, seven.

Hoffman.—Not an uncommon species throughout the wooded areas of the northern part of the State.

Ridgway.—Found along the entire route from Sacramento eastward, but only in certain widely separated localities.

Walla Walla. J. W. Williams.—May 20, 1885, first seen, about thirty; seen every day afterward; common May 20–23; still present August 9; not very common summer resident; young shot in July.

79. **Melanerpes uropygialis** (Baird). GILA WOODPECKER.

Heermann.—Found in considerable numbers on the Colorado.

Cooper, 1870.—At Fort Mojave I found this woodpecker abundant in winter. About March 25 they were preparing their nests in burrows near the dead tree tops.

80. **Colaptes auratus** (Linn.) FLICKER.

Forrest Ball (Auk. October, 1885). San Bernardino, January, 1885, a female specimen; identification approved by Mr. Robert Ridgway.

[A male with a black mustache would have been more satisfactory evidence of the occurrence here of this species. *Colaptes auratus hybridus* (Baird), has been dropped from the A. O. U. list without explanation, probably because it is still a puzzle unsolved. The many specimens in various plumages collected on this coast deserve at least a passing notice].

Contra Costa and Alameda counties. W. E. Bryant.—Specimens of *Colaptes* hitherto referred to *hybridus* are now taken almost as often as *C. cafer;* in fact, it is unusual to get really good examples of *C. cafer* in some localities.

Haywards. W. O. Emerson.—Three seen all winter.

[Mr. C. A. Allen some years ago informed me that he had taken a fine series at Nicasio. Dr. Cooper, 1870, refers to an Oakland specimen as differing from *auratus* only in having the head grayish like *mexicanus* and the black cheek feathers tipped with red. I have never seen a specimen on the Pacific Coast which had any black in the mustache, nor have I been able to get anything but the typical *mexicanus* in the mountains of Central California and incline to the opinion that most of the mixed individuals termed hybrids are but winter visitants to California, and that the most remarkable of these are adults in the most perfect plumage. I collected a fine series at Marysville, winter of 1878, in freezing weather.]

Sebastopol. F. H. Holmes.—Collected here (no date).

Ukiah. G. E. Aull.—Winter visitant.

Camp Harney. Bendire.—A well marked specimen shot by Lieutenant D. Cornman in the spring of 1875.

81. Colaptes cafer (Gmel.) Red-shafted Flicker.

San Diego. L. B.—Common. No "hybrids" noticed in northern Lower California or about San Diego. It probably does not meet *chrysoides* in northwestern Lower California, but is likely to do so on the Colorado desert.

Poway. F. E. Blaisdell.—Common resident. Volcan Mountains, common August 21.

Volcan Mountains. W. O. Emerson.—First seen February 25; March 1, common, and rapping on dead limbs and calling.

San Bernardino Mountains. F. Stephens.—Rare; foothills, less rare; common in the valleys where it breeds. Agua Caliente, San Diego County, west end of Colorado desert, several seen; probably resident in small numbers.

Henshaw, 1876. Found in southern California without reference to special locality, being common both in the mountains and low districts.

Santa Cruz. Joseph Skirm.—Common; after breeding it frequents orchards and feeds on fruit. A clutch of eggs is five or six, rarely seven.

Contra Costa and Alameda counties. W. E. Bryant.—Common resident.

Central California. L. B.—Very common resident and generally distributed.

Willamette Valley. O. B. Johnson.—Abundant; nesting commonly.

Beaverton. A. W. Anthony.—Common resident; slight increase in numbers March 1, 1884.

Cooper, 1860. Constant resident in Washington Territory, at least west of the Cascades.

Suckley, 1860. Extremely common in the timbered districts of Washington Territory.

British Columbia. John Fannin.—Abundant resident.

Henshaw, 1879. An abundant, widely distributed species. The birds of the eastern slope appear to be typical *mexicanus* and I have never seen a specimen from this region showing intermediate characters.

Camp Harney. Bendire.—Very common, arriving the latter part of March.

Hoffman. Common.

Ridgway. Being the most abundant and generally distributed of the woodpeckers, this species was found in all wooded localities.

82. **Colaptes cafer saturatior** Ridgw. NORTHWESTERN FLICKER.

[I have been unable to find this form in the Upper Sacramento Valley, but Mr. Charles H. Townsend collected it at Red Bluff in winter and in the redwood forests of Humboldt County, at the latter in winter also, I believe.]

83. **Colaptes chrysoides** (Malh.) GILDED FLICKER.

Cooper, 1870. I found only two pairs of this species at Fort Mojave after Feb. 20.

Heermann. In considerable numbers on the Colorado.

84. **Phalænoptilus nuttalli californicus** Ridgw. CALIFORNIA POOR-WILL.

Agua Caliente, San Diego County. F. Stephens.— Present on my arrival (March 15, 1886.) Heard nightly; rather common.

Poway. F. E. Blaisdell.—A summer resident; seems to be confined to mountains, foothills and small cañons, not appearing in open valleys.

Poway. W. O. Emerson.—April 20, 1884.

Henshaw, 1876. On the summits of the mountains near Fort Tejon, remarkably numerous.

Santa Cruz. Joseph Skirm.—Very rare in this vicinity. I have seen but five individuals since I came here. Mr. A. M. Ingersoll found the eggs in 1883; they were on the bare ground; color, pure white.

Cooper, 1870. I have neither heard nor seen any west of the Coast Range nor in the Santa Clara Valley in spring. They are, however, common in the hot interior valleys and remain near San Francisco as late as November. Dr. Kennerly obtained one in the Colorado Valley February 23, 1854.

Contra Costa and Alameda counties. W. E. Bryant. Rare summer resident.

Berkeley. T. S. Palmer.—Tolerably common transient visitant. Last seen October 29, 1885.

Nicasio. C. A. Allen.—Common in summer; first seen March 1, 1884; male shot.

Sebastopol. F. H. Holmes.

Central California. L. B.—Very rare in summer; more common in the foothills in spring and fall, and an uncertain number winter in them. At North American Hotel, 30 miles east of Stockton, January 22, 1879, I flushed one several times but did not get it, and the landlord, Mr. Stroud, told me of a whippoorwill his laborers found a week previous while clearing off some land by the hotel—land that had a dense growth of chapparal (*Adenostoma fasciculatum*) on it. The bird, he said, was torpid, and he took it up and looked at it. The bird I flushed was on a grass-covered hill side where the sun was shining, and it flew as well as usual, though going only a few yards at a time. I have seen a few here at other times in February and March, got a specimen February 22, 1889, and shot one at Gridley, October 28, 1885, probably a recent arrival, as a pine needle was sticking to its feathers. At Lake Tahoe, September 8, 1889, I saw two, and heard these, or others, several evenings in August and September. A fine specimen was shot at Copperopolis, February 19, 1886, while flying, by Mr. Andrew Simpson, and mounted by Dr. Davenport of Stockton.

Newberry. This species is found in all parts of California and Oregon. On the shores of Rhett Lake we came upon its nest.

Henshaw. It arrives at Carson from the south in the early days of May. It soon becomes generally and commonly distributed over nearly all the region embraced in the present report, being scarcely less numerous towards the north.

Camp Harney. Bendire.—Rare summer visitor.

Ridgway. An inhabitant of open places exclusively, the sage brush country being, so far as we observed, its only habitat. It was often observed or heard in the lower valleys, as well as in the mountain parks below an altitude of 8,000 feet (Truckee, East Humboldt Mountains, Uintahs.)

Hoffman. I met with the bird several times; the first locality being in the valley west of Hot Spring Cañon, on the road to Belmont. They appeared rather frequently in the vicinity of Green Mountain and north of Mount Magruder.

85. Chordeiles virginianus henryi (Cass.) WESTERN NIGHTHAWK.

Poway. F. E. Blaisdell.—Summer resident; March 21, 1884, first.

Poway. W. O. Emerson.—April 23, 1884, first.

[Probably a part, if not all of the above, were *C. texensis*].

Henshaw, 1876. Appears to be wanting in much of southern California; we did not meet with the species at all.

Cooper, 1870. This species shuns the coast borders of the State. In Santa Clara Valley and the Coast Ranges I have seen none.

L. B.—A very common summer resident in the granite rocks at about 7,000 feet altitude in the Sierra, as far south as Tuolumne and Alpine counties. It was common at Butte Creek House, altitude 5,800 feet, July 1–3, 1884—about as low as I have found the species breeding. It was last seen at Summit, September 1, 1885; Lake Tahoe, September 11, 1889. It probably winters in Lower California but I have not collected it there; supposed I saw one in April near the Cape.

Chico. Wm. Proud.—June 17, 1884, first seen; rare.

O. B. Johnson. Common in summer, breeding on gravelly islands in the Willamette River.

Cooper, 1860. Very abundant in the interior of the Territory, arriving at Puget Sound about June 1, remaining until September.

Suckley, 1860. Abundant at Fort Dalles and on the prairies near Puget Sound. At Fort Steilacoom, arrived June 1, 1854; June 3, 1856.

British Columbia. John Fannin.—Common summer resident.

Henshaw, 1879. An extremely abundant summer visitant through California, Oregon, and Washington Territory.

Camp Harney. Bendire.—An exceedingly common summer resident, arriving about May 20 and leaving early in October.

Ridgway. A common summer inhabitant of the country traversed.

Hoffman. Found south of Eureka, on the northern slopes of Prospect Hill.

[Like Dr. Cooper I think it shuns the coast, at least south of San Francisco, and I have never seen it in the San Joaquin or Sacramento valleys.]

Beaverton. A. W. Anthony.—June 1, 1885, one bird, first; breeds.

Burrard Inlet. John Fannin.—May 26, 1885, first; May 28 next; June 15 common; breeds.

86. Chordeiles texensis Lawr. TEXAN NIGHTHAWK.

San Diego to San Pedro Mountain, May 9–20, tolerably common the entire route. Quite common on the *mesa* about San Diego in summer. April 8, 1884, several, first seen.—L. B.

San Bernardino Valley. F. Stephens.—Rare summer resident.

Cooper, 1870.—On the 17th of April I saw the first of this species at Fort Mojave. About the 25th of May they were paired. I found them as far west as the Coast Mountains.

Baird, Brewer and Ridgway, vol. iii, 521. Dr. Cooper shot a single specimen of this species near San Buenaventura, April 18, 1873.

87. **Cypseloides niger** (Gmel.) BLACK SWIFT.

San Diego, May 21, 1881, a flock of twenty or more flew over the *mesa* near town. As it was sunset when they passed, I supposed they were going to their nests for the night. The next evening I stationed myself where I had seen them, and got a specimen, as they flew in the same direction they did the previous evening. Big Trees, June 8, three seen; Weber Lake, August 1, 1889, one seen. Seen on a few other occasions in mountains of Central California in late summer.—L. B.

Seattle. O. B. Johnson.—April 9, 1884.

British Columbia. John Fannin.—Tolerably abundant summer resident. Burrard's Inlet, May 26, 1885, first; May 28, next seen; common June 15; breeds.

Ridgway.—Truckee Reservation, May 31, 1868, specimen. On the 23d of June following we found it abundant in the valley of the Carson River. They were evidently breeding in the locality.

Haywards. W. O. Emerson.—April 19, 1885, thirteen birds.

88. **Chætura vauxii** (Towns.) VAUX'S SWIFT.

San Diego, April 28, 1884, two specimens from a small flock which was seen but a short time during a cool rain storm.

Campo, May 14; showery; a small number.

San Diego, April 16, 1885, first seen, three; April 29, a flock; also about twenty-five in Santa Margarita cañon, sixty miles north of San Diego, April 26, 1885. A single bird seen 125 miles southeast of San Diego May 16, 1885.—L. B.

San Bernardino. F. Stephens.—Rare transient visitant to the valley.

Baird, Brewer and Ridgway (vol. iii, p. 521).—Dr. Cooper states that in the spring of 1873, this swift appeared as early as April 22 near San Buenaventura. The year before he first saw them near San Diego on the 26th.

Nicasio. C. A. Allen.—May 12, 1884, first seen and shot.

Sebastopol. F. H. Holmes.—An irregularly common summer resident. April 19, 1885, first seen; next seen April 29.

Beaverton. A. W. Anthony.—1884, seen several times this season. April 30, 1885, six seen; next seen May 8; common May 8; breeds.

British Columbia. John Fannin.—Summer resident; not common.

Ridgway. At the Truckee Reservation, near Pyramid Lake, in May and June, 1868, we saw nearly every evening, but never until after sundown, quite a number of small swifts which must have been this species, but they always flew at so great a height that we found it impossible to obtain a specimen.

Murphys. John J. Snyder.—September 4, 1885, a specimen.

89. **Micropus melanoleucus** (Baird). WHITE-THROATED SWIFT.

Cooper, 1870. About twelve miles north of San Diego,

rather numerous about some rocky bluffs close to the sea shore, March 22. I saw none at Fort Mojave until May. On the 7th of June, near the head of the Mojave River, I found a few about some lofty granite cliffs.

San Diego, December 15, 1883, a few near the bay. El Cajon, twelve miles east of San Diego, January 16, 1884, a hundred or more flying about a pond, several shot; all very fat, proving that insect food was ample, in winter, here, for their support. I heard of these afterward in January, but swollen unbridged streams and very bad roads restricted my observations in February, March and April to the immediate vicinity of San Diego, and consequently I did not again look for them, and did not again see the species at San Diego.—L. B.

San Bernardino. F. Stephens.—Tolerably common summer resident of the foothills; rare winter visitant to the valley and foothills. Agua Caliente, March 25–28, 1884, common in neighboring cañons. In 1886, seen from March 24 to April 15 (day of leaving).

Baird, Brewer and Ridgway.—Dr. Cooper saw many of this species in the cañon of Santa Ana, May 20. He also saw them near San Buenaventura, August 25, when they came down the valley from the sandstone cliffs ten miles distant. They afterwards hunted insects daily near the coast, flying high during the calm morning, but when there were sea breezes flying low and against it. After a month they disappeared and none were seen until December 14, when they were again seen until the 20th. None were seen during the rains or until February 26, when they reappeared. After April 5 they retired to the mountains.

Port Harford. L. B.—March 29, 1881, two or three dozen. Summit, Central Pacific Railroad, July 2, 1885 from four to six dozen, also seen in August. Breeds in lava cliffs of Calaveras County.

Haywards. W. O. Emerson.—January 25, 1885, ten or more have been about town off and on for a week.

Contra Costa County. W. E. Bryant.—Rare summer resident.

Ridgway. First noticed in the early part of July, 1868, on the Toyabe Mountains near Austin—a single bird only. At the Ruby Mountains a little later in the same month we found it extremely numerous about the high limestone cliffs. At this place they literally swarmed.

Hoffman. Rather common in the more elevated regions, building in and about the fissures and projections of cliffs; was noticed again in the upper portion of the Black Cañon of the Colorado valley in September.

90. Trochilus alexandri Bourc. & Muls. BLACK-CHINNED HUMMINGBIRD.

San Diego. L. B.—Rare migrant; perhaps breeding; if so overlooked.

Poway. F. E. Blaisdell.—Common summer resident; began to breed in April.

Poway. W. O. Emerson.—April 21, 1884, young.

San Bernardino. F. Stephens.—Common; breeds in the valleys.

Agua Caliente, San Diego County, F. Stephens.— Ten specimens seen from March 19 to April 13, 1886.

Henshaw, 1876. Not found by our party very common in any portion of California.

Cooper, 1870. Mojave River, June 3, breeding; several nests near Santa Barbara in April and early in May.

Central California. L. B.—Apparently rare or local, in summer only, seen occasionally in the Sierra at varying heights, usually in August and September, common in breeding season along the San Joaquin and Sacramento Rivers at certain localities. Winters entirely south of California though I did not see it in the Cape region.

Chico. Wm. Proud.—April 19, 1884, first; bulk arrived April 21,

British Columbia. John Fannin.—Summer resident; not common.

Henshaw, 1879. By no means a common summer resident along the eastern slope and seems to have there a limited distribution; specimens procured in June along Honey Lake; present in July near Camp Bidwell.

Hoffman. Found in the valleys of the northern interior of Nevada.

Ridgway. The only hummer which was encountered along every portion of the route, in the proper localities, it being equally common at Sacramento and among the mountains of Utah. Truckee Valley, May and June abundant.

Marysville. W. F. Peacock —April 4, 1885, first; again April 5; common May 12; breeds.

91. **Trochilus violajugulum** Jeffries. VIOLET-THROATED HUMMINGBIRD.

Known only from a single specimen from Santa Barbara, Cal.

92. **Trochilus costæ** (Bourc.) COSTA'S HUMMINGBIRD.

San Diego. N. S. Goss.—March 17, 1884, first male.

San Diego. L. B.—March 21, first female; weather fine for several days on and previous to the 17th; mercury about 70° from sunrise to sunset; *Calandrinia Menziesi*, *Anagallis arvensis* and *Nemophila aurita*, about in full flower. Winters wholly south of San Diego, where it is a common summer resident.

San Bernardino. F. Stephens.—Tolerably common

summer resident; breeds in the cañons of the foothills.

Colorado Desert. F. Stephens.—March, 1886. Young of the year. Rather common, and breeding at my arrival (March 18). Resident as a species.

Henshaw, 1876.—None were detected by our parties. [Mr. Henshaw was at Santa Barbara and vicinity the most of the time from June 1 to July 13, during that time visiting Los Angeles, and the fact that neither of the naturalists of the party or parties met with *costæ* sufficiently indicates its rarity so far north. Mr. Evermann, Auk, 1886, page 179, says " I have but one specimen obtained in Ventura County." He collected two years in that county.]

Cooper, 1870. At San Diego, in the backward spring of 1862, I first saw them April 22, and have since found them north to San Francisco, where, however, they are rare. I did not observe any at Fort Mojave until March 5, and they were not numerous afterward.

93. Trochilus anna (Less.) ANNA'S HUMMINGBIRD.

San Diego to San Pedro Mountain, May 9-20, 1885, tolerably common on the entire route. San Diego, tolerably common summer resident; more common in winter.—L. B.

San Diego. B. F. Goss.—March 8, 1884, juveniles, two-thirds grown.

Poway. F. E. Blaisdell.—Common summer resident.

San Jose. A. L. Parkhurst.—Mating February 4, 1885.

Volcan Mountains. W. O. Emerson.—March 11, a bright warm morning, with a light layer of snow on the ground, a hummingbird darted by me on its way down the cañon. From its size, color, dark throat and head, it could be no other at this altitude (6,000 feet) than the hardy little *anna*.

Haywards. W. O. Emerson.—I found a nest January 19, 1886, my earliest record.

San Bernardino. F. Stephens.—Common resident of valley and foothills. Agua Caliente, common; probably resident in neighboring cañons.

Henshaw, 1876. During the summer we saw none in the low valleys, but found it reasonably numerous in the mountains.

Santa Cruz. Joseph Skirm.—Quite rare summer resident.

Alameda and Contra Costa counties. W. E. Bryant.—Common resident.

Berkeley. T. S. Palmer.—Abundant resident.

A. M. Ingersoll. Mr. H. R. Taylor found eggs February 14, 1885; incubation far advanced.

Nicasio. C. A. Allen.—February 21, 1884, first male.

Central California. L. B.—Tolerably common, especially in the foothills, where a few winter. Gridley, December 1, 1885 (November mild), a pair daily during November among the flowers in the garden.

94. Trochilus floresii (Gould). FLORESI'S HUMMINGBIRD.

The second known specimen was recorded from San Francisco, by W. E. Bryant, Forest and Stream, xxvi, 426.

95. Trochilus platycercus Swains. BROAD-TAILED HUMMINGBIRD.

Ridgway. We first encountered the broad-tailed hummingbird on the Rocky Mountains, where it was very abundant in July and August.

96. Trochilus rufus Gmel. RUFOUS HUMMINGBIRD.

San Diego, March 10, 1884, first males; both sexes common a few days later. In 1885 not a *Selasphorus* was seen from March 23 to May 9; migrating land birds were much rarer than they were in the spring of 1884, probably owing to scantier vegetation, fewer insects and much less rainfall, the latter the principal cause of rarity. Not seen by me in any part of Lower Califorfornia though no doubt wintering abundantly there.— L. B.

San Diego. Cooper, 1870.—First arrived February 5, 1862; several seen February 22.

Poway. F. E. Blaisdell.—Only two individuals seen.

Poway. W. O. Emerson.—Young April 21, 1884.

Haywards. W. O. Emerson.—April 6, 1885. The first arrived in 1886 on February 16.

San Bernardino. F. Stephens.—(Including *T. alleni*) from rare to tolerably common transient visitant in valley and foothills. This species or *T. alleni* was seen at Agua Caliente March 25–28. In 1886 was seen March 18, 19; April 7–13.

Henshaw, 1876. Quite common in summer and breeds apparently as commonly in the valleys as in the mountains.

Ventura County. B. W. Evermann (Auk, 1886.)— This I consider the most abundant species of the hummers found in the county. It is resident except for a few weeks in midwinter.

Marysville. W. F. Peacock.—Tolerably common summer resident. First arrival March 23; common, April 19, 1885.

Alameda and Contra Costa counties. W. E. Bryant. Tolerably common summer resident.

Chico. William Proud.—One March 17, 1885.

Sebastopol. F. H. Holmes.—Common summer resident. First, April 1; common, April 15, 1885.

Berkeley. T. S. Palmer.—Common summer resident. First arrival, female, February 6, 1885; male,

February 11; nest of two eggs April 2. In 1886, first, February 19; common, March 25. A nest of this species containing three eggs was found May 20.

Central California. L. B.—Rare summer resident of Tuolumne, Calaveras and Alpine counties above 4,000 feet; common July 1-2 north end of Butte County; rather common migrant through the entire region and really abundant in August and September in the upper Sierra. Last seen September 9, 1885, at Summit, having all left during cool weather and first frost. September 21, one seen Summit 1886. I have collected a great deal in the Sacramento Valley from Marysville and vicinity northward, but have not seen it in either the Sacramento or San Joaquin Valley in breeding time.

Nicasio. C. A. Allen.—March 21, 1884, first seen and shot.

Willamette Valley, O. B. Johnson.—Common summer resident, breeding.

Beaverton. A. W. Anthony.—Common summer resident; March 28, 1884, first, not again seen for some days; April 15 increasing rapidly, common, April 21. May 1, first nest. In 1885, first arrived March 27; common April 3; first female April 2.

Suckley, 1860. Abundant in western portions of Oregon and Washington Territories and Vancouver Island.

Cooper, 1860. Very abundant in Washington Territory reaching the Straits of Juan de Fuca as early as March 17, 1884, when I saw them in considerable number. They all leave the Territory in September.

British Columbia. John Fannin.—Very abundant summer resident. Burrard Inlet, arrived April 5; common May 4, 1885.

Henshaw. Probably rather local as a summer resident of the eastern slope. It appeared to be rather numerous at Camp Bidwell in summer.

Camp Harney. Bendire.—Rare; very few seen.

Ridgway. In the rich valley of the Lower Truckee the only species of hummingbird found in August at which time great numbers were seen. Same locality in May and June of the following season not one of this species was found, its place being taken, apparently, by *T. alexandri*. We next saw the rufous-backed hummer in the West Humboldt Mountains. Eastward seen only in Secret Valley the most eastward point to which it is known to extend.

97. **Trochilus alleni** (Hensh.) ALLEN'S HUMMINGBIRD.

San Diego. Arrived about the same time as *T. rufus*, apparently rare, probably not always distinguished from it in the field; male and female identified by specimens shot by Mr. Carl H. Danielson and myself.—L. B.

Santa Cruz. Joseph Skirm.—Quite rare summer resident.

Sebastopol. F. H. Holmes.—Rare April 15, 1885.

Oakland and vicinity. W. E. Bryant.—Tolerably common summer resident.

Haywards. W. O. Emerson.—Common summer resident. Arrived February 15, 1885,

Nicasio. C. A. Allen.—March 13, 1884, first.

Olema. A. M. Ingersoll.—April 4, 1884, nest, and young about a week old. I have also found it breeding at Santa Cruz.

L. B.—Not yet detected in the central part of California; see Bull. Nutt. Orn. Club, July, 1877, October, 1877, for description of this bird; also January, 1878.

Mr. Charles W. Gunn, (O. & O. February, 1885,) who collected at Colton, San Bernardino County, says of it: "Not common; five specimens taken, found in company with *T. rufus*."

98. **Trochilus calliope** Gould. CALLIOPE HUMMING-BIRD.

Volcan Mountains. N. S. Goss.—April 15, 1884, male and female.

San Bernardino Mountains. F. Stephens.—Rare; breeds in the pine region.

Agua Caliente, San Diego, Cal. F. Stephens.—One specimen, April 13, 1886.

Henshaw, 1876. A single individual in the Tejon Mountains, August 17; unaccountably rare in the mountains of southern California.

Berkeley. T. S. Palmer.—Rare accidental visitant.

Central California. L. B.—Rare migrant through valleys and foothills; breeds in the Sierra above 4,000 feet; rather rare at Big Trees in breeding season; common in north Butte County at this time; not seen in Lower California.

British Columbia. John Fannin.—Burrard Inlet common summer resident. Arrived May 7; common May 28, 1885.

Ridgway. Ruby and East Humboldt Mountains, at an altitude of 7,500–10,000 feet, abundant in August and September.

Henshaw, 1879. Its summer habitat appears in general to be limited by the eastern slope along which in Nevada, California and Oregon it was found by our parties to be very numerous.

99. **Tyrannus tyrannus** (Linn.) KINGBIRD.

Ridgway. In the valley of the Truckee River two or more pairs had their abode among the large cottonwood trees near our camp, July 24, August 28.

Bendire. In the John Day River valley and sixty miles to the east at the Malheur agency, very common. I have not seen it at Camp Harney.

Walla Walla. J. W. Williams.—This bird was first seen here May 20, 1885; six individuals noticed, and afterward every day. It became common May 25. Young seen in June and July; August 19, (date of report), still present.

Suckley, 1860. I found it at Puget Sound where I obtained several skins. Among the cottonwood trees fringing the lakes of Nisqually Plains, August 5, 1853, I obtained a nest containing nearly fledged young.

British Columbia. John Fannin.—Very rare summer resident.

100. Tyrannus verticalis Say. ARKANSAS KINGBIRD.

Winters entirely south of California, though not detected by me in the Cape region, nor does Xantus appear to have found it there. I found a pair at San Quintin Bay, May 8, 1881, and shot one of them. Very few appear to breed south of San Diego, as it was rare at Campo and southward in May, 1884. San Diego, March 25, first—a male specimen; April 1, both sexes common.— L. B.

Poway. F. E. Blaisdell. — Summer resident; first eggs, May 27; last seen, August 20, 1884.

Santa Isabel. W. O. Emerson.—April 3, 1884, many pairs.

Julian. N. S. Goss.—April 5, 1884.

San Bernardino. F. Stephens.—Common summer resident of the valley; rare summer resident of the foothills. Agua Caliente, several seen. In 1886, first seen March 26; quite common from April 1 to April 15.

[Bakersfield, March 27, 1889, and common between Stockton and Bakersfield, March 26.]

Sebastopol. F. H. Holmes.—First seen April 13; common May 1, 1885; common summer resident.

San Jose. A. L. Parkhurst.—April 8, 1884, first—one male; breeds; rare in 1885; arrived March 22.

Nicasio. C. A. Allen.—April 8, 1884, first; April 12, 1876, first.

Berkeley. T. S. Palmer.—Rare summer resident. Last seen August 29, 1885. Arrived May 11, 1886.

Olema. A. M. Ingersoll.—April 22, 1884, first.

Haywards. W. O. Emerson.—April 12, 1885, two males and a female.

Stockton. L. B.—March 26, 1879, first; March 31, common. Common in central California in summer, below about 3,000 feet, as it no doubt is in all the principal agricultural districts of California.

Stockton. J. J. Snyder.—First seen March 29, 1885; soon became common.

Murphys. J. P. Snyder.—April 16, 1885; common April 26; last seen August 31. Young male shot October 17, 1884. Common summer resident.

Marysville. W. F. Peacock.—Common summer resident. April 16, first; male and female; bulk arrived April 21, 1884. In 1885 arrived March 31; common April 12.

Gridley. L. B.—April 3, 1890, eleven or twelve seen.

Chico. Wm. Proud.—April 12, 1884, first, a few passed over; April 16 eleven birds came into the garden in the evening; not seen next morning. In 1885 arrived April 6, one bird.

Beaverton. A. W. Anthony.—First seen May 30; again June 1, 1885. Very rare.

Willamette Valley. O. B. Johnson.—Common in summer, breeding.

Walla Walla, W. T. Dr. Williams.—April 26, 1885, fifty specimens; common May 27, still present August 9.

Cooper, 1860. Arrives at Puget Sound in June together with the common species. They never approach the coast.

Suckley, 1860.—Fort Dalles about May 15, 1853.

Yakima, W. T. Samuel Hubbard, Sr.—Common summer resident. First seen May 15, 1885.

British Columbia. John Fannin.—Rare summer resident.

Henshaw, 1876. Very numerous summer resident.

Bendire.—Common summer resident; generally distributed; arrive at Camp Harney about May 1.

Ridgway.—Generally distributed throughout all fertile districts of the west.

Hoffman.—Found breeding in June south of the Central Pacific Railroad, in the valleys between Austin, Hot Spring Cañon and Belmont, in the cottonwood groves.

101. **Tyrannus vociferans** Swains. Cassin's Kingbird.

San Diego, common summer resident; from rare to tolerably common in San Diego County near sea level in December and January, 1883-84, but was not seen from January 29 to March 7, during which time there was great rainfall and much chilly weather.—L. B.

Cooper, 1870.—Resident as far north as Los Angeles during winter.

Poway. F. E. Blaisdell.—Summer resident, less common in winter.

Poway. W. O. Emerson.—January and April, 1884.

San Bernardino. F. Stephens.—Rare summer visitant to the valley.

Henshaw, 1876. Los Angeles, June; not seen elsewhere.

[Mr. B. W. Evermann informed me in 1880, that he had collected it at Santa Paula, Ventura County, and that it was more common there than *T. verticalis*. Dr. Cooper (Cal. Orn.) states that he found some of these birds in Santa Clara Valley in May, 1864, which were smaller and greener on the back than those from the south. He also says they winter in small numbers at

Santa Cruz. My impression is that they are now seldom found so far north at any time.]

102. Myiarchus cinerascens Lawr. ASH-THROATED FLYCATCHER.

San Diego.—Common summer resident; April 9, 1884, first; wind strong from south-southeast. April 19, common.—L. B.

Poway. F. E. Blaisdell.—Common summer resident; May 31, first eggs; last seen August 18, 1884. In 1885 the first bird arrived April 9.

Julian. N. S. Goss.—April 22, 1884.

San Bernardino. F. Stephens.—Tolerably common resident of the valley and foothills.

Agua Caliente, San Diego County. F. Stephens.—April 6–15, 1886.

Henshaw.—Generally distributed over the southern portion of the State, and common; avoiding the heavy timber and mountains.

Tehachapi. L. B.—First seen April 5, 1889; common two days later.

Santa Cruz. Joseph Skirm.—Common summer resident.

San Jose. A. L. Parkhurst.—April 20, 1884, first, two or three.

Alameda and Contra Costa counties. W. E. Bryant. Tolerably common summer resident.

Berkeley. T. S. Palmer.—Rare summer resident. First seen April 30, male and female; next seen May 19, 1885. In 1886 the first arrived May 9; next seen May 15.

Nicasio. C. A. Allen.—May 1, 1884, first; April 27, 1876, first.

Haywards. W. O. Emerson.—Rare summer resident. First seen, April 25, 1885; arrived in pairs.

Stockton. L. B.—April 14, 1879, first; common April 22. A common summer resident in Central California to about 3,000 feet altitude.

Stockton. J. J. Snyder.—Arrived April 12, seven birds; again April 19, 1885.

Marysville. W. F. Peacock.—Common summer resident. April 24, 1884, first male; common May 14. In 1885, the first arrival April 12; common April 13.

Sebastopol. F. H. Holmes.—Common summer resident. April 9, 1885; common April 13.

Henshaw. Common on the foothills near Carson; apparently rare farther north. A single individual was seen at Honey Lake, California.

Camp Harney. Bendire.—A rare summer visitor frequenting the juniper groves and breeding in deserted woodpeckers' holes.

Ridgway. A few were observed among the cottonwoods of the Lower Truckee in July and August; also not an infrequent summer resident in the cañons of the Ruby Mountains. Carson River, June 24, 1868, seem to be breeding.

Cooper, 1870. I found one at Fort Mojave January 15, and think a few may habitually winter in the Colorado Valley.

Jacksonville, Oregon. W. E. Bryant.—Breeding season of 1883. [Mr. Bryant's is the most northern west slope record.]

103. Sayornis saya (Bonap.) SAY'S PHŒBE.

San Diego, either one or two pairs nesting in April, 1885. Never numerous in California; most common in winter.—L. B.

Poway. F. E. Blaisdell.—Winter visitant; last seen March 8, 1884.

San Bernardino. F. Stephens.—Tolerably common

winter visitant to the valley. Rare summer resident, breeding in the same.

Santa Ana. F. E. Blaisdell.—December 10-14, 1884, frequently seen.

Santa Cruz. Joseph Skirm.—Quite common in fall and winter.

Contra Costa County. W. E. Bryant.—Tolerably common resident.

Sebastopol. F. H. Holmes.—Rather rare winter visitant. September 17, 1884, first; February 24, 1855, last.

British Columbia. John Fannin. — Rare summer resident.

Henshaw. Abundant along the eastern slope, building about barns and outbuildings.

Camp Harney. Bendire.—Rare and only found during the migrations, usually about April 1.

Ridgway. Eastward of the Sierra Nevada, found in all suitable places, but not abundant anywhere. Its favorite haunts were the rocky shores of the lakes and rivers, or the walls of the lower cañons in the mountains, but wherever man has fixed his abode upon the dreary waste this species was attracted to his vicinity.

104. **Sayornis nigricans** (Swains.) BLACK PHŒBE.

Campo, January and May, rare; San Diego, moderately common resident; Murphys and Colfax, winter as well as summer, remaining through snow storms; Red Bluff, February 3-5, 1885, rare, a common but never numerous species, breeding occasionally at Big Trees in outbuildings; rarely found so high on the west slope. A single bird was seen at Summit, August 26, 1885, coming from the northeast where it had probably spent the summer; altitude of Summit 7,000 feet. A nest at Dunbar's, Calaveras County, altitude 3,700 feet was composed mostly of soap root fibres with the usual covering of mud.—L. B.

Poway. F. E. Blaisdell.—Breed.

San Bernardino. F. Stephens.—Tolerably common resident of the valley. Agua Caliente, common; probably resident. March 18 to April 15, 1886, two or three daily.

Henshaw, 1876. Quite numerous in California in the southern portion.

Alameda and Contra Costa counties. W. E. Bryant. Common resident.

Nicasio. C. A. Allen.—March 28, 1876, first.

Newberry. Common in Northern California; specimens obtained in Umpqua Valley, Oregon.

Willamette Valley. O. B. Johnson.—A single example July, 1879.

Ridgway. Found only at Sacramento.

Henshaw, 1879. Appears not to be present in this region. [East slope.]

Sierraville, Sierra Valley. June, 1885, a pair.—L. B.

Cooper, 1870. An abundant and resident species in all the lower parts of California except the Colorado valley where I saw none later than March 25, 1861, as they had gone north.

105. Contopus borealis (Swains.) OLIVE-SIDED FLYCATCHER.

Not found in the Cape region, but winters entirely south of California. Hansen's, 60 miles south of Campo, May 10–12 (1884), at 6,000 feet altitude, common; Tia Juana, April 30, 1885, a single migrant shot; a rare migrant through the low parts of California.— L. B.

Agua Caliente. F. Stephens.—One, April 7, 1886.

Poway. F. E. Blaisdell.—April 23, 1884, first; Volcan Mountains, August.

Julian. N. S. Goss.—April 21, 1884, first.

Henshaw, 1876. In southern California appears not to be as numerous as in the middle region, pretty closely confined to the mountains.

Cooper, 1870. Dr. Gambel found young at Monterey in July; rather common in the Coast Range toward Santa Cruz where they had nests in May.

L. B.—Common in the fir forest of the Sierra of Central California on both slopes, in Calaveras and Butte counties on the west, Alpine and Sierra on the east; arriving rather late in spring and starting south from the first to the middle of September. Its nests are usually forty or fifty feet from the ground, rarely as low as twenty, in cone bearing trees, and are mostly composed of yellow lichen *(Evernia vulpina)* lined sparingly, in several instances, with fine, wiry rootlets.

Beaverton. A. W. Anthony.—Common summer resident. May 11, first; bulk arrived May 22, 1884.

Willamette Valley. O. B. Johnson. Common in summer.

Cooper, 1860. Very common arriving early in May; remains until late in September.

British Columbia. John Fannin.—Rare summer resident.

Henshaw, 1879. Occurs in summer all along the eastern slope up to the Columbia River, and probably still farther north. It does not appear to be as numerous here as in the Rocky Mountains, or even in the region west of the Sierra.

Camp Harney. Bendire.—A very rare summer visitor.

East Humboldt Mountains. Ridgway.—The first individual.

Hoffman. A common summer resident confined to the more elevated coniferous regions, at least as far south as Belmont.

Beaverton. A. W. Anthony.—May 5, 1885, first; May 6 next seen; common May 10. Tolerably common; breeds.

106. Contopus richardsonii (Swains.) WESTERN WOOD PEWEE.

First seen at San Diego, April 28, 1884, one male. In 1885, first seen April 20. I first saw it at Stockton in 1880 on May 2. At Gridley, April 30, 1886; the first male, Stockton, May 7, 1889.—L. B.

Poway. F. E. Blaisdell. — Summer resident; first seen April 29, 1884.

Julian. N. S. Goss.—May 1, 1884.

San Bernardino. F. Stevens.—Rare summer resident of the foothills. Agua Caliente, one April 7 and two April 11, 1886.

Berkeley. T. S. Palmer.—Last seen September 12, 1885.

San Jose. A. L. Parkhurst.—April 28, 1884, first.

Haywards. W. O. Emerson.—Rare summer resident. Arrived May 9, 1885.

Alameda and Contra Costa counties. W. E. Bryant. Tolerably common summer resident.

Marysville. W. F. Peacock. — Tolerably common summer resident. Arrived May 14, 1885; common June 6.

Sebastopol. F. H. Holmes.—Abundant summer resident. Arrived April 21; common April 24, 1885.

Olema. A. M. Ingersoll.—May 17, 1884, first.

Nicasio. C. A. Allen.—May 11, 1884, first.

Beaverton. A. W. Anthony.—Common summer resident. Common May 15, 1885.

Willamette Valley. O. B. Johnson.—Very common in summer.

British Columbia. John Fannin.—Summer resident, not common. (Burrard Inlet. Arrived May 24, 1885.)

Walla Walla. J. W. Williams.—Summer resident.

Henshaw, 1879. Common summer resident of the mountains.

Camp Harney. Bendire.—Moderately common summer visitor.

Hoffman. Common throughout the northern and more timbered regions of Nevada but rather rare in the southern interior.

Ridgway. Met with in every wooded locality and was no less common at an altitude of 8,000 feet in the Wahsatch Mountains than at Sacramento, but little above sea level.

L. B.—It almost always places its nest on a dead horizontal limb, at least this is according to my observations and I have seen many nests which were saddled on limbs, in a solitary instance, however, the nest was in or on horizontal diverging twigs in a deciduous oak where it was partly hidden by foliage; again, one was nicely surrounded and to a great extent concealed by having been built in a bunch of yellow lichen (*Evernia*.) Stockton, September 15, 1883, is my latest record.

107. Empidonax difficilis Baird. WESTERN FLYCATCHER.

San Diego, April 12, 1884, first; several males; April 20 first female. In 1885 the first arrival was March 20. I think a few remain in San Diego County ordinary winters; I have seen one there in December.—L. B.

Poway. F. E. Blaisdell.—April 15, 1884, first. A summer resident. In 1885, first arrival March 17.

San Bernardino. F. Stephens.—Tolerably common summer resident of the valleys and mountains.

Agua Caliente. F. Stephens.—Seen from April 2 to 15, 1886; common the latter part of the time.

Henshaw, 1876. Not uncommon summer resident in

southern California. They spend the summer from sea level up to 7,000 feet, but are most numerous in the mountains.

Santa Cruz. Joseph Skirm.—Breeds.

Berkeley. T. S. Palmer.—Tolerably common summer resident. First seen April 7; common April 13, 1886.

San Jose. A. L. Parkhurst.—Common summer resident. Arrived March 22, six specimens; next seen March 24, 1885. (Three specimens September 18, 1886, near San Jose.—W. O. Emerson).

Alameda and Contra Costa counties. W. E. Bryant. Common summer resident.

Haywards. W. O. Emerson.—Common summer resident. The first flycatchers to arrive, March 21, 1885; common April 1.

Nicasio. C. A. Allen.—April 2, 1884, first; April 6, 1876, first.

Olema. A. M. Ingersoll.—April 7, 1884, first.

Central California. L. B.—Breeds sparingly from the valleys to near summits of Sierra. Gridley, April 30, 1886, one male shot.

108. Empidonax acadicus (Gmel.) ACADIAN FLYCATCHER.

Ridgway. The rarest of the *Empidonaces*, a few being seen in the pine forests high up on the Wahsatch Mountains and a still smaller number on the eastern slope of the Sierra Nevada.

British Columbia. John Fannin.—Common summer resident.

Burrard Inlet. John Fannin.—May 26, 1885, first; May 28 next; common June 6; breeds.

109. **Empidonax pusillus** (Swains.) LITTLE FLY-CATCHER.

Campo, May 7-9 rare.—L. B.

Poway. F. E. Blaisdell.—May 6, first.

San Bernardino. F. Stephens.—Rare migrant in valley and foothills.

Henshaw, 1876. Abundant in southern California, especially so in the swampy thickets about Los Angeles. Specimens at Los Angeles and Santa Barbara in June; Fort Tejon and Tejon Mountains in August.

L. B.—Very common summer resident in willows of Central California, most so along the valley rivers; eggs taken at Blood's, altitude 7,200 feet; breeds at many localities in the Sierra, as at Summit; Butte Creek House; Sierra Valley and at Hermit and Hope valleys farther south; arrives at Stockton about May 1 (April 30, 1878, May 4, 1880, the latter a backward spring); May 7, 1889, two shot; at this date several seen but all were silent.

Beaverton. A. W. Anthony.—May 22, first; common June 7; in the last week in July I found two nests and eggs on large ferns.

O. B. Johnson, 1880. Quite common in summer.

Walla Walla. Dr. Williams.—June 16, 1885, six miles from the post (identification correct).

British Columbia. John Fannin.—Common summer resident.

Burrard Inlet, B. C. John Fannin.—May 26, 1885, first; May 28 next; common June 6; breeds.

Henshaw. Numerous summer resident well up into Oregon.

Bendire. I saw a number May 8, 1876, on Rattlesnake Creek.

Ridgway. The most abundant and generally distributed of the *Empidonaces*. Specimens at Sacramento, Ruby Valley and other localities.

Hoffman. Specially abundant along the Humboldt River and tributaries from the north.

Cooper, 1870. At Fort Mohave on the first of May I found several of them inhabiting a very dark dense thicket, being attracted by their note which sounded like queat-queah. I afterward heard their peculiar notes along the Mojave River, near Los Angeles, and in May, 1863, at Santa Barbara.

[Mr. Ridgway says the people of Parley's Park translated their notes as "pretty dear." This and Dr. Cooper's "queet queah" represent its notes very well—notes which differ much from those of the other small flycatchers of the Pacific Coast. Although looking considerable alike in the field, with specimens for comparison there is no difficulty in separating them, not even the immature of *E. hammondi* and *F. wrightii*, the bill of *E. hammondi* being much the smaller, that alone is sufficient to distinguish them; *E. hammondi* has also a shorter tarsus and is a smaller, frailer bird. Drs. Cooper and Suckley reported *E. pusillus* as common and abundant at Puget Sound and Fort Steilacoom, and this is the only small fly catcher mentioned by them, but they probably also saw others which they confounded with *E. pusillus*, which probably never "flits through the upper branches of the tall spruces" nor are its notes "short but sweet, particularly low, plaintive and soothing," as stated by Dr. Suckley].

110. **Empidonax hammondi** (Xantus). HAMMOND'S FLYCATCHER.

San Diego, April 26, 1884, first male; few seen here; none southward, although it winters entirely south of California.—L. B.

Agua Caliente. F. Stephens.—One April 14, 1886.

Poway. F. E. Blaisdell.—April 14, 1884, first (identification correct).

Julian. N. S. Goss.—April 15, 1884.

Henshaw, 1876. I could find no evidence that it breeds in southern California; after September common in the mountains; remains till into October.

Central California. L. B.—By no means rare during migrations. Instead of being a common summer resident of the pine forests of the Sierra as I stated in Proc. Nat. Mus., 1879, I now think it a rare summer resident. The only nest I have found was on June 6, 1880, at Big Trees. This was on a horizontal limb of a living pine, forty or fifty feet from the ground, and partly hidden by foliage. It was very wide in proportion to its depth. By shooting it down the eggs were destroyed. The female was shot as she flew from the nest and was sent to the Smithsonian Institution. A few remain in this latitude as late as September 15, arriving from the south about May 1.

Henshaw, 1879. Said to occur along the eastern slope but I did not meet with it.

Ridgway. Not met with anywhere as a summer resident but during its autumnal migrations was very common on the East Humboldt Mountains.

Baird, Brewer and Ridgway. Mr. Dall found it breeding in Alaska. Eggs unspotted, creamy white.

[The few notices of the species must not be taken as proof of its rarity].

111. Empidonax wrightii Baird. WRIGHT'S FLYCATCHER.

It is very rare in north Lower California about the middle of May, arriving at San Diego April 20, 1884; at Stockton May 1, 1879, April 30, 1880; mostly going south by September 1, but a few are at latitude 39° to about the 15th. Their nests found by me in Central California were in *manzanita* and other shrubs. It is a common summer resident in the fir forests, a common migrant through all parts of Central California, not breeding below the firs.—L. B.

Poway. F. E. Blaisdell.—Noticed on several occasions.

San Bernardino. F. Stephens.—Rare migrant through the valley.

Agua Caliente, San Diego County. F. Stephens.—One seen April 8, 1886.

Henshaw, 1876. A few near Mt. Whitney in September.

Ridgway. As characteristic of the mountains as *E. pusillus* is of the lower valleys. Common in May in the Lower Truckee Valley; first observed near Carson on the 21st of April. It was equally common on both sides of the Great Basin, the only districts where entirely absent being those where the ranges were destitute of water and vegetation.

Henshaw, 1879. Apparently rather uncommon as a summer resident of the mountains. A nest found June 22 on an open bush contained four fresh eggs, yellowish white, unspotted.

Camp Harney. Bendire.—A single specimen taken May 15, 1884.

Butte Creek House. L. B.—Latitude 40° 10′, July 1–2, rare.

Baird, Brewer and Ridgway (vol. iii, p. 520). Dr. Cooper found a few of this species wintering in a large grove of balsam, poplars and willows, which retained most of their old leaves till spring, near San Buenaventura. Those shot were remarkably gray, and were supposed to have been blown down from the borders of the desert by the violent northeast wind.

112. **Pyrocephalus rubineus mexicanus** (Scl.) VERMILION FLYCATCHER.

F. E. Blaisdell. I examined a specimen while at Santa Ana December 10-14, 1884, which was killed December 9. At that time it was said to be rather common along the Santa Ana River.

Baird, Brewer and Ridgway (vol. iii, p. 520). Dr. Cooper found two male birds of this species in a grove near the mouth of the Santa Clara River six miles from Santa Buenaventura in October, 1872.

Heermann, 1854. I had the good fortune to procure at Fort Yuma a specimen of this flycatcher, which Dr. Milhau informs me is there quite common in spring.

113. **Otocoris alpestris leucolæma** (Coues). PALLID HORNED LARK.

114. **Otocoris alpestris chrysolæma** (Wagl.) MEXICAN HORNED LARK.

This form of the horned lark is abundant from San Diego to Sonoma County, and also occurs in San Joaquin County. Young of the year able to fly were already common April 20, 1884, in San Diego.—L. B.

Poway. F. E. Blaisdell.—Common resident.

Volcan Mountains. W. O. Emerson.—February 24, 1884, large flocks feeding on the open flats with lark finches; large numbers in Santa Isabel, January 23.

Alameda and Contra Costa counties. W. E. Bryant.—Common resident.

115. **Otocoris alpestris rubea** Hensh. RUDDY HORNED LARK.

This is the resident form of *Otocoris* of the Sacramento Valley, and probably of the greater part, if not all, of the San Joaquin Valley.

I have shot good examples of it in winter and early summer in San Joaquin, Stanislaus and Calaveras counties, and have shot specimens in Kern County that I referred to it, though none of these Kern County specimens were as richly colored as the best examples from the Sacramento Valley. It is common as far north as Red Bluff in winters of average mildness. There is great individual variation in this as in the other forms of *Otocoris* which occur in California. I have collected specimens of *rubea, strigata, chrysolæma* and Mr. Dwight's recently described *merrilli* at or near Stockton, and in Yuba, Butte and Sutter counties have collected *rubea, strigata* and *merrilli*, and think specimens which might be referred to *chrysolæma* can easily be obtained in the three last named counties in summer and winter, but believe that *strigata* and *merrilli* are but winter visitants to the valleys of California. For valuable articles on the perplexing varieties of the horned larks see Henshaw, Auk, vol. i, 254, and Dwight, Auk., vol. vii, 138.

Leucolæma has been collected as far south as Carson, in winter, by Mr. Ridgway.

116. Otocoris alpestris strigata Hensh. STREAKED HORNED LARK.

Gridley, October 9, 1884, first seen; a large flock. In about a week it became common and remained all winter; was still at Gridley March 8, when I left it, but I could not find it here after March 23, 1890.

Mr. A. L. Parkhurst, of San Jose, collected it there and sent me a specimen for identification—the most southern point where it has been taken so far as I know, but it probably goes much farther south in severe winters. It was in large flocks at Red Bluff, February 2, 1885. It is very common at and about Stockton in winter.

Summit, August 16, 1885; in cool weather for the time of the year, a flock of thirty or forty arrived and remained several days. They were mostly young birds in very dark spotted plumage, very different from the young Californian bird, which is much paler. Adults of both sexes were shot and positively identified.

This is the form which Professor Ridgway referred to typical *alpestris* in Birds of Central California, Proceedings Nat. Museum, 1879. I presume that Mr. Anthony's bird belongs here, and Dr. Suckley's also, though I know nothing of their specimens.—L. B.

Beaverton. A. W. Anthony.—Large flocks arrived April 17, 1885; a few stayed until April 30, after which I saw none.

Suckley, 1860. A very abundant summer resident on the gravelly prairies near Fort Steilacoom.

117. Pica pica hudsonica (Sab.) AMERICAN MAGPIE.

British Columbia. John Fannin. — Abundant resident. I have never seen the yellow-billed magpie here (1884).

Suckley, 1860. On Puget Sound not observed until August, 1856, after which time during the fall they became moderately abundant. I have never observed the yellow-billed magpie in Oregon.

O. B. Johnson. Quite common in the vicinity of Forest Grove; probably breeds.

Fort Klamath. Lieutenant Wittich.—A common species. May 12, 1878, nest with four young in a thorny bush or low tree.

Sierra Valley. L. B.—June, 1885, several. Summit, November 16, 1884, a single bird; same locality, September 30, 1875, one about a slaughter-house; probably a straggler from the east slope. Alpine County, on the east slope, several seen; said to be a resident up to 7,000 feet; probably never occurs in California west of the Sierra.

Henshaw, 1879. In the settled portions of Nevada and eastern California the magpie is a constant resident.

Camp Harney. Bendire.—Not common during the summer; more abundant throughout the winter.

Ridgway. In western Nevada from the Sierra eastward to the West Humboldt Mountains it was one of the most abundant species. It was abundant in the rich valleys of the Truckee and Carson rivers.

118. Pica nuttalli Aud. YELLOW-BILLED MAGPIE.

Cooper, 1870. This magpie is abundant in the valleys of California, especially near the middle of the State. At Santa Barbara I found them numerous in April and May, and saw their nests in oak trees, but the young were nearly fledged by the 25th of April. They breed abundantly about Monterey. Their food consists of almost everything, animal and vegetable.

Henshaw, 1876. In the Sierra proper we did not meet with these birds, but in various parts near the sea coast they were very numerous.

Oakland. W. E. Bryant.—One bird seen in winter, probably an escaped one.

Central California. L. B.—Common resident; rare at Red Bluff in winter. It is becoming scarce about Stockton and some other towns. Colonies breed in certain localities a long time if not molested. It rarely gets above about 1,200 feet in the foothills of the Sierra and probably does more good than harm, though it is likely to be exterminated because of its fondness for the eggs of domestic fowls.

Burrard Inlet. John Fannin.—1885, a single individual.

119. Cyanocitta stelleri (Gmel.) STELLER'S JAY.

Inhabits the Coast region from northwest California to Sitka.

120. Cyanocitta stelleri frontalis Ridgw. BLUE-FRONTED JAY.

Not found in mountains south of Campo. Tehachapi, common.—L. B.

F. E. Blaisdell. Common in June in pine regions of San Diego County.

Volcan Mountains. W. O. Emerson.—In firs and cedars; winter.

San Bernardino Mountains. F. Stephens.—Common resident.

Henshaw, 1876. A common inhabitant of the mountains, rarely being seen in summer below 5,000 feet.

Cooper, 1870. Inhabiting the Coast Range as far south as Santa Cruz at least.

Alameda and Contra Costa counties. W. E. Bryant.— Rare resident.

L. B.—Lower foothills of central California, and borders of the principal valleys, in winter, moderately common; a very common summer resident in the fir forests; rare at Donner Lake and Summit, November 12, 1884; wintering chiefly between two thousand and four thousand feet altitude, usually nesting high in conifers, a single nest seen as low as six feet in a small *libocedrus*. At Blue Cañon Mr. A. M. Ingersoll found eight nests inside of the snow sheds, May, 1886.

Chico. Wm. Proud.—Numerous all winter (1884–'85). I had not seen them here before in eight years. The last was seen June 5 in a cañon a few miles east of Chico.

Willamette Valley. O. B. Johnson.—Abundant resident; nesting in communities.

Cooper, 1860. Very common in all the forests of the Territory both sides of the Cascades.

British Columbia. John Fannin.—Very abundant resident.

Henshaw, 1879. This form of Steller's jay has been traced by the expedition from the Coast and Sierra ranges of southern California into the Cascade mountains of Oregon, and so on up to the Columbia River, at which point, however, the form does not cease, but continues into Washington Territory.

Camp Harney.—Bendire.—A rare resident.

Ridgway. We found this jay only among the pines on the Sierra Nevada.

121. **Aphelocoma woodhousei** (Baird). WOODHOUSE'S JAY.

Ridgway.—At our camp on the western slope of the Humboldt Mountains it was very abundant in September. In Buena Vista Cañon it was also common; also rather common on the eastern slope of the Ruby Range.

122. **Aphelocoma californica** (Vig.) CALIFORNIA JAY.

San Diego. L. B.—Common resident. Campo, January, common.

Poway. F. E. Blaisdell.—Common resident; begins to breed about the last of April.

San Bernardino. F. Stephens.—Rare resident of the valley and foothills. Agua Caliente, foothills, probably resident.

Henshaw, 1876. Found on the mountains to a height of about 5,000 feet; farther up than which it begins to be rare.

Alameda and Contra Costa counties. W. E. Bryant. Common resident.

L. B.—Common in most of the agricultural districts of California; rarely seen above 3,500 feet in the Sierra, latitude 38°. Common at Red Bluff, February, 1885;

much more numerous in Butte County winter of 1884 and 1885 than usual, probably in part winter visitants in search of food or a milder climate.

Wilbur, Oregon. W. E. Bryant.—Breeds.

Willamette Valley. O. B. Johnson.—Common among deciduous trees, breeding about habitations.

Nuttall. Near Fort Vancouver, early in October.

Henshaw, 1879. Numerous in the foothills to a considerable distance north of Carson. A specimen was taken at the Dalles, October 4. Mr. C. Roop of Portland, Oregon, informed me that this jay is abundant near the mouth of the Columbia River, both in Oregon and Washington Territory.

Ridgway. On the east slope it appeared to be quite common, at least on the foothills near Carson, where in 1868 it made its first appearance toward the last of April.

L. B.—I occasionally see individuals on the east slope in autumn, which appear to be migrants on their way to California to spend the winter. A few have been seen at Lake Tahoe and the Summit of the Central Pacific Railroad late in September and early October.

123. **Aphelocoma insularis** Hensh. SANTA CRUZ ISLAND JAY.

124. **Perisoreus obscurus** (Ridgw.) OREGON JAY.

Newberry. In California we found them at the upper end of the Sacramento Valley, in latitude 40°. As we progressed toward the Cascades it became more common.

John Feilner, Sm. Rept. 1864. This bird I first saw in 1859 about Lassen's Butte. About Shasta Butte, on the north and east side, May 15. I found them in large numbers, up to twenty together, noiselessly and busily engaged searching for insects on the ground.

Willamette Valley. O. B. Johnson.—Common in heavy timber in winter.

Beaverton. A. W. Anthony.—Common February 2, when I arrived here. It appears to breed early, after which it disappears. I took a nest March 31; the species was last seen April 15.

Cooper, 1860. Mouth of the Columbia River, March 1854, a small scattered flock industriously seeking insects and seed among the spruce trees much in the manner of the titmice. I have seen a few at Puget Sound.

British Columbia. John Fannin.—Very abundant resident, widely distributed over the province. On the Arctic slope I have seen these birds pure white.

Henshaw, 1879. Resident of the mountains from northern California to the Columbia River; young in nesting plumage taken near Camp Bidwell. Along the upper Des Chutes River the "meat birds," as they are suggestively termed, were very numerous.

125. Corvus corax sinuatus (Wagl.) AMERICAN RAVEN.

San Diego. Common resident.—L. B.

San Bernardino. F. Stephens.—Tolerably common resident of the valley. Agua Caliente. A pair seen, probably resident.

Cooper, 1870. Found in pairs everywhere in California and the adjacent regions. I obtained numbers at Fort Mojave.

Tehachapi. L. B.—Very common here and vicinity, April and March, 1889.

Henshaw, 1876. I saw ravens on Santa Cruz Island.

Santa Cruz. Joseph Skirm.—Rare.

Sebastopol. F. H. Holmes.

L. B.—Decidedly rare in Central California since I began to collect and pay special attention to birds, about 1876. I know of but two or three pairs in this part of the State since then, and these were seen only in winter. A few have been seen going from Nevada to California in fall.

Newberry. A constant feature in all parts of the country traversed.

O. B. Johnson, 1880. Not rare in the vicinity of Forest Grove.

Cooper, 1860. On the barren arid plains east of the Cascades they were very common, while the common crow was rarely seen. At Vancouver, however, in winter, I observed them amicably associating together.

British Columbia. John Fannin.—Very abundant resident.

Henshaw, 1879. More or less common everywhere save in the higher mountains.

Camp Harney. Bendire.—A common resident.

Hoffman. Everywhere more or less abundant, and a permanent resident. [Nevada.]

Ridgway. One of the most characteristic species of the Great Basin over which it appears to be universally distributed. We did not see it in the Sacramento Valley.

126. Corvus americanus Aud. AMERICAN CROW.

San Diego. Common resident; breeds in cañons and valleys a few miles from the coast, probably very rare in Lower California and only in the extreme north.— L. B.

Poway. F. E. Blaisdell.—Common in favorable localities; breeds here. Volcan Mountains, August 21 to November 28, common.

Volcan Mountains. W. O. Emerson.—Seen occasionally in winter. There was a large rookery in a valley at the base of the mountain.

San Bernardino. F. Stephens.—Common resident of the valley.

Cooper, 1870. In the southern half of California the crow is rarely seen on the sea beach but prefers the inland districts, occasionally, however, coming to the shores of bays to feed. I never saw one in the Colorado Valley nor near the summits of the Sierra Nevada.

Santa Cruz. Joseph Skirm.—Rare.

Alameda and Contra Costa counties. W. E. Bryant.—None seen for several years; it formerly bred at Berkeley.

Central California. L. B.—Abundant resident in the valleys; not seen in the Sierra in several years; no record of seeing it there since 1878.

Fort Klamath. Lieutenant Wittich.—Rarely seen in this region.

Willamette Valley. O. B. Johnson.—Common resident, breeding in communities.

Cooper, 1860. Near the coast it appears rarely, the fish crow replacing it.

Suckley, 1860. Occasionally met with in Washington Territory but it is not abundant, being replaced by the succeeding [*Corvus caurinus.*]

Henshaw, 1879. In Oregon, especially east of the mountains, crows are very numerous.

Camp Harney.

Ridgway. East of the Sierra Nevada so extremely rare as to be met with on but two occasions. Specimens at Truckee Meadows November 8 and Humboldt Meadows October 31.

Hoffman. Along the cliffs a few miles southeast of Bull Run Mountain, and again in a similar locality, at the southern extremity of Snaky Valley, found in considerable numbers. Specimens were seen in California near Partzwick, just across the line.

[I have never been able to detect any difference in the notes of crows of California, Nebraska, or Central Pennsylvania, and doubt if there is any].

127. Corvus caurinus Baird. NORTHWEST CROW.

British Columbia. John Fannin.—Very abundant resident east and west of the Cascades; winters on the coast; is smaller than *C. americanus*. In breeding season it utters a note not unlike the coarse mewing of a domestic cat.

128. Picicorvus columbianus (Wils.) CLARKE'S NUT-CRACKER.

British Columbia. John Fannin.—Summer resident east of the Cascades.

Cooper, 1860. Quite abundant along the banks of the Yakima River, whence it continued common northward. Appeared at Vancouver during the severe cold winter of January, 1854, in considerable numbers. I have never seen it at any other season west of the Cascades.

O. B. Johnson, 1880. Common in the Cascade Mountains down to the foothills in winter.

Fort Klamath. Lieutenant Wittich.—They sometimes come about the quarters, and kept about the kitchens.

Henshaw, 1879. A constant and abundant resident of the pineries.

Camp Harney. Bendire.—Moderately common during the winter and spring months; none during the rest of the year; breeds very early in the season. On the 5th of May, 1875, I found young birds well able to fly. "Both nests were in pine trees; one was on the extremity of a branch about twenty-five feet from the ground."

Ridgway. On the Sierra Nevada most abundant; found to the eastward wherever extensive coniferous woods occurred.

Hoffman. The only locality where seen during the whole journey was on the northern slope of Mt. Nagle.

Central California. L. B.—Common in summer from

about 6,000 feet upward; rare at Summit and Donner Lake November 12-16; not at Big Trees January, 1879. Abundant on the Summit (latitude 39° 15') in autumn of 1885, invariably moving southward. This migration began about August 15. Mr. Ridgway (Bull. Essex Inst.) found it abundant at Carson, thirty or forty miles distant, in the winter of 1867-68. They frequently come about kitchens for the refuse, and mountaineers are compelled to enclose carcasses of sheep in sacks in order to prevent these birds from eating the meat.

Blood's. W. E. Bryant.—June 25, 1885, two caught alive in a log storehouse, which they probably entered for the meat hanging there.

Henshaw, 1876. During September met with in great numbers.

Tahachapi. L. B.—Moderately common in the pines.

San Bernardino Mountains. F. Stevens.—Rare resident or winter visitant.

129. **Cyanocephalus cyanocephalus** (Wied). PIÑON JAY.

Cooper, 1870. I am told by Mr. Clarence King that they frequent the junipers on mountains near Mariposa.

Summit. L. B. — November 14, 1884, one bird; September 25, 1885, one specimen, which was alone. September 29, about twenty following the divide southward. September 30, two specimens. Lake Tahoe, September, 1889; numerous sojourner.

Newberry. We first noticed this bird in the Des Chutes Basin, latitude 44° 12', in September, every morning flocks of from twenty-five to thirty, etc.; fifty miles farther north they were feeding on the berries of the cedar *(Juniperus occidentalis)*.

Henshaw, 1879. At certain points in the Des Chutes Basin it was noticed in great numbers.

John Fielner. Immediate vicinity of Shasta Butte, May 15, seen in numbers. The first time I saw this bird was at Fort Tejon. (Sm. Report, 1864.)

Camp Harney. Bendire. — Late in October, 1875, about eighty; a flock flying southward.

Ridgway. On the 21st of April full grown young were flying about in a cedar and piñon grove near Carson. In this grove we found the abandoned nests, perhaps a hundred or more. These nests were saddled upon the horizontal branches, at a height of eight or ten feet from the ground. Mr. Ridgway (Bull. Essex Inst.) says they were abundant at Carson in winter.

Hoffman. Rather common throughout that belt of coniferous trees represented by the piñon*(Pinus edulis)*, extending more particularly and uninterruptedly from Eureka southward to below Hot Spring Cañon. Again, at the occurrence of the same belt at Mount Nagle and Mount Magruder, the species was detected, but not so frequently.

130. **Dolichonyx oryzivorus albinucha** Ridgw. WESTERN BOBOLINK.

Ridgway. We found it common in August in the wheat fields at the Overland Ranch in Ruby Valley, Nevada, and we were informed at Salt Lake City that it was a common species on the meadows of that section of the country in May, and again in the latter part of summer when the grain ripened.

131. **Molothrus ater** (Bodd.) COWBIRD.

British Columbia. John Fannin. — Rare summer resident.

Ridgway. We found this species so rare in the country traversed that the list of specimens given below comprises every individual seen during the whole time.

Truckee Reservation, June 2, 1868, male adult. Valley of the Humboldt, August 31, 1867, two juveniles. Utah, June, 1869, two eggs.

Hoffman. Several specimens were seen at Camp Independence, Cal., in August but nowhere else during the expedition.

Cooper, 1870. They seemed to be migrating northward through the Colorado Valley early in April, and on the 19th of that month I found an egg in a nest of the yellow-breasted chat, showing that some of them are raised in the latitude of 35° as well as northward.

Heermann. I remarked a flock of these birds as far south as Fort Tejon.

[Variety *obscurus* winters on the Gulf of California at Guaymas, and perhaps breeds in the Colorado Valley or still farther north.]

132. **Molothrus ater obscurus** (Gmel.) DWARF COWBIRD.

I found it common in the extreme southern part of Lower California, where Xantus also found it in 1859, but have not seen it elsewhere in Lower California and no cowbirds have ever been collected in California, west of the Sierra Nevada, as far as I am aware. Nuttall, Edition 1840, p. 191, says the cow blackbird exists in California.—L. B.

133. **Xanthocephalus xanthocephalus** (Bonap.) YELLOW-HEADED BLACKBIRD.

San Diego, irregular winter visitant. April 19, 1884, a flock of about 1,000 in which were but three or four in female plumage.—L. B.

Temecula. F. E. Blaisdell.—May 4, 1883, abundant.

San Bernardino. F. Stephens.—Very rare winter visitant; rare transient visitor in the valley.

Agua Caliente. Several seen near the spring, probably winter visitants; several seen March 18 and 21, 1886.

Cooper, 1870. They build at Santa Barbara and northward, avoiding the immediate coast.

Fresno. Gustav Eisen.—Rare here this winter (1884); abundant in Tulare County.

Northern San Joaquin and Sacramento Valleys. L. B.—Abundant summer resident, breeding in the extensive tule marshes. A few may be found about Stockton during the very mildest winters. Last seen at Gridley, October 6, 1884, one in female plumage.

Newberry. We found them with young at Pit River, and immense flocks swarmed in the rushes bordering the Klamath Lakes.

Fort Klamath. Lieutenant Wittich.—Summer resident.

Henshaw, 1879. Abundant in all suitable localities, along the eastern slope far towards the Columbia River.

Camp Harney. Bendire.—A very common species breeding abundantly amongst the tules in Malheur Lake; none remain during the winter.

Ridgway. It was abundant in the vicinity of Sacramento and along the southeastern margin of Salt Lake; it was also plentiful at all intermediate points where suitable localities existed. The species was partly migratory, only a few examples being seen during the winter at Carson.

Hoffman. Common in all the marshy regions and grassy meadows from Independence Valley southward to Fort Mojave, excepting in the southern interior valleys which appear too much isolated and restricted.

134. Agelaius phœniceus (Linn.) RED-WINGED BLACKBIRD.

British Columbia. John Fannin.—Common resident.

Cooper, 1860. During the winter at Vancouver remaining in small flocks.

Seattle. O. B. Johnson.—February 16, 1884, a flock of nine; February 12, cold; skating on Lake Union.

Seattle, February 17, snowstorm; snow 18 inches deep.

Seattle, February 29, first peeping frogs heard.

Willamette Valley. O. B. Johnson, 1880.—Very abundant in summer; breeding.

Fort Klamath. Lieutenant Wittich.—Nesting in great numbers.

Henshaw, 1879. An abundant summer resident along the eastern slope; more or less winter about Carson.

Camp Harney. Bendire.—A very abundant summer visitor; some remain through the winter.

Hoffman. Common in nearly all the marshy districts, in the northern and middle regions, being found in company with the yellow-headed blackbirds.

Ridgway. Found in all the marshy places, being especially numerous in the vicinity of the great lakes of the interior, and along the large rivers. Specimens at Truckee Reservation, May 15 and May 31.

Central California. L. B.—Abundant in winter. Gridley, October 1, 1884, first, a large flock in female plumage. Stockton, October 6, 1881, first, a flock. Stockton, October 1, 1883, first, a flock in female plumage only. San Diego, rare winter visitant; April 24, 1884, last. Southern Lower California, near Cape St. Lucas, rare in winter.

Walla Walla. J. W. Williams.—Winters here.

135. **Agelaius gubernator** (Wagl.) BICOLORED BLACKBIRD.

San Diego, common resident.—L. B.

San Bernardino. F. Stephens.—Common summer resident of the valley.

Cooper, 1870. I found them in scattered pairs in May throughout the Coast Range even to the summits, where there are small marshes full of rushes in which they build. According to my observations this species inhabits chiefly the interior of the State, Santa Cruz being the only point on the coast where I have seen them.

Santa Cruz. Joseph Skirm.—Common resident.

W. E. Bryant. Tolerably common summer resident of Alameda and Contra Costa counties.

Berkeley. T. S. Palmer.—Rare summer resident.

Central California. L. B.—Abundant resident in the valleys; common in winter to near the head of the Sacramento Valley. The bird of Sierra Valley has the notes of *gubernator* and the wing patch of *phœniceus*; the notes of these nearly allied forms may be alike, but my impression is that they differ.

Dr. Cooper (Cal. Orn., 1870), says he has been unable to detect any difference in their notes and habits. He found a nest in the Coast Range which was formed of grass and rushes, lined with finer grass, which agrees nearly with several I have found, though Dr. Heermann says the nest is composed of mud and roots, lined with fine grass.

Fort Klamath. Lieutenant Wittich. — Nesting in great numbers.

Willamette Valley. O. B. Johnson.—Very abundant, with habits similar to *phœniceus*.

Henshaw. This form is less abundant than *phœniceus*, as well as less generally distributed along the eastern slope. It is noticeable in this connection that none of the specimens which are referable to this form, from localities east of the mountains, represent it in its typical condition.

Camp Harney. Bendire.—An abundant species, but less abundant than *A. phœniceus*. The eggs of these two species present but little difference. As a general thing the eggs of this form are not marked so much, and a few are unspotted.

Ridgway. East of the Sierra Nevada, found only in the western depression of the Great Basin (Western Nevada), and was there very rare compared with *A. phœniceus*. Specimen, Carson, March 9, 1868. Nest and eggs, Truckee Reservation, June 3.

Stockton. L. B.—April, 8, 1879, first nest and full set of eggs; young able to fly May 20, 1878.

136. Agelaius tricolor (Nutt.) TRICOLORED BLACKBIRD.

San Diego. Common winter visitant and resident in suitable localities.—L. B.

Bernardino River. F. E. Blaisdell.—A colony breeding May 16, 1883; May 25 all had hatched, four birds in each nest. The young seemed to be of about the same age.

San Bernardino. F. Stephens.—Common; breeds in the valley.

Cooper, 1870.—I found them the most abundant species near San Diego and Los Angeles, and not rare at Santa Barbara.

Tehachapi. L. B.—April 6 and 7, flocks going to San Joaquin Valley from the Mojave Desert.

Henshaw, 1876. I found the species breeding in but one locality, in Santa Clara Valley, June 21.

Santa Cruz. Joseph Skirm.—Common summer resident.

Oakland and vicinity. W. E. Bryant.— Rare winter visitant.

Sebastopol. F. H. Holmes.

Newberry. Common in California, in the Klamath Basin and Oregon.

Central California. L. B.—Abundant summer resident; rare as far north as Stockton in mildest winters; quite common in January, 1885. An immense colony was breeding here May 10, 1879, in the tules. The nests were attached to the tules of this and last year's growth; were composed wholly of grass without mud, differing in this respect from nests found in northern California by Dr. Heermann. The greatest number of eggs or young birds in any one of about 100 nests examined was three. Several held but one egg or one bird; several young were dead and the colony was not thriving. Many of the parents were going back and forth in small parties, bringing grasshoppers for their young from a pasture about three miles distant. The nests averaged about one to each square yard.

137. **Sturnella magna neglecta** (Aud.) WESTERN MEADOWLARK.

British Columbia. John Fannin. Abundant resident.

Cape Beale, B. C. Emanuel Cox.—Always here.

Cooper, 1860. Very abundant in all the prairies of the Territory where it resides constantly, merely visiting the warmer coast meadows in very cold weather but not remaining there during summer.

Yakima Valley. Samuel Hubbard, Jr.—February 17, 1885, one; March 1, common; breeds.

Beaverton. A. W. Anthony.—Tolerably common February 2, more plentiful March 15, 1884. A great many were frozen and drifted under in the snow storm of December 14.

Portland. M. F. Spencer.—February 19, 1884, first song heard.

Willamette Valley. O. B. Johnson.—Constant resident, less common in winter.

Central California. L. B.—Abundant resident below the fir forests; rare in them, in summer only.

Berkeley. T. S. Palmer.—Abundant resident.

Alameda and Contra Costa counties. W. E. Bryant. Common resident.

Henshaw, 1879. Found in the fertile valleys and on the plains.

San Bernardino. F. Stephens.—Common resident of the valley; abundant winter visitant of the same; Agua Caliente, several, probably wintering. Seen from March 18 to April 15, 1888.

Volcan Mountains. W. O. Emerson.—Seen but once. Santa Isabel, common and singing at all times.

Poway. F. E. Blaisdell.—Common resident; begins to breed in March.

San Diego. L. B. — Abundant resident; nesting March 15. One of the few really abundant species of the agricultural districts of California and apparently so of the most of those of the entire Pacific coast.

Henshaw, 1879. Very numerous throughout this whole region.

Camp Harney. Bendire.—A very abundant summer visitor, breeding everywhere in low lands as well as in the highest mountain meadows. About Camp Harney they raise two or three broods in a season.

Ridgway. A generally distributed species; it is much less common in the mountains, however, than in the lower valleys. Carson, abundant in winter, in sage brush and fields.

Hoffman. The grassy valleys are the usual resort, although the adjacent prairies are also visited, especially the northern portion of Nevada, where the undulating country is more abundantly covered with Compositæ, the southern region being either bare of vegetation or presenting *Artemisia*, etc., in excess.

138. Icterus parisorum Bonap. SCOTT'S ORIOLE.

Appropriately named by Lower Californians, Oriole of the Mountains, or its equivalent in their language; found also down to near sea level.

I found several males about fifteen miles south of the boundary line May 6, 1884, and still nearer the line May 9, 1885. Mr. F. Stephens informs he that he collected seven specimens at Campo in 1882. I also noticed it near Campo in May, and between Campo and Hansen's. Mr. Stephens appears to be the only one who has taken it north of the line, though I saw a pair fifteen miles east of San Diego May 16, 1884.

Dr. Cooper says he saw a bird at Fort Mojave in April which he supposed to be this; he could hardly be mistaken.

139. Icterus cucullatus nelsoni Ridgw. ARIZONA HOODED ORIOLE.

Tia Juana, near boundary line. N. S. Goss.—March 21, 1884, three males.

San Diego. L. B.—March 30, 1884, first males.

Poway. F. E. Blaisdell.—March 22, 1884, males; April 2, females; eggs May 22; last seen September 20. First seen March 11, 1885, male; April 4 first female.

San Bernardino. F. Stephens.—Tolerably common summer resident of foothills; breeding here. Agua Caliente, seen March 29, 30, April 6, 14 and 15, 1886.

Poway. F. E. Blaisdell.—March 11, 1885, first male.

L. B.—San Diego to San Pedro Mountain via San Rafael, tolerably common May 9–10, 1885; not above about 2,500 feet.

Santa Paula, Ventura County. B. W. Evermann.—A common summer resident. I have traced it as far north as Santa Barbara.

[Not found by Mr. Henshaw in 1876 at Los Angeles, Santa Barbara, etc., or if so, not mentioned.]

Cooper, 1870. I found this species arriving at San Diego about April 22, and they were not rare for a fortnight afterwards, but then retired into the warmer interior valleys, where I have seen them nearly as far north as Los Angeles.

140. Icterus bullocki (Swains.) BULLOCK'S ORIOLE.

San Diego. L. B.—Rare summer resident. March 22, 1884, first males.

Poway. F. E. Blaisdell.—Summer resident; March 21, 1884, first males; eggs taken May 9; last seen August 19. First seen March 17, male; April 3, 1885, female.

Julian. N. S. Goss.—April 11, 1884, a male; April 17, a female.

San Bernardino. F. Stephens.—Common summer resident of the valley; rare summer resident of the foothills. Agua Caliente, March 25–28, common and migrating. Seen every day from March 18 to April 15, 1885.

Los Angeles. Henshaw.—June, 1876, very common. Fort Tejon, August 27.

Bakersfield. L. B.—March 27, 1889, males. Tehachapi, April 6, 1889, first female.

Grayson. W. E. Bryant.—March 24, 1884.

Santa Cruz. Joseph Skirm.—April 3, 1881, first; April 17, 1882; April 16, 1883.

Haywards. W. O. Emerson.—Common summer resident. First seen April 10, two males; common April 15, 1885.

San Jose. A. L. Parkhurst.—April 3, 1884, first; two

or three males in song. December 5, a male (unusually late). First seen in 1885, March 23, a few, singing. Common summer resident.

Alameda and Contra Costa Counties. W. E. Bryant. Common summer resident. (H. R. Taylor, Alameda. March 20, 1885, two males.)

Berkeley. T. S. Palmer.—Tolerably common summer resident. First seen March 20, 1885. In 1886, first seen March 22, one male; again March 25; common May 11.

Nicasio. C. A. Allen. — April 20, 1884, first; April 19, 1876.

Olema. A. M. Ingersoll,—April 17, 1884, first.

Stockton. Dr. E. C. Davenport, April 2, 1884, first.

Stockton. J. J. Snyder.—First seen March 23; common April 5, 1885. Females usually arrive about a week later than the males.

Stockton. L. B.—March 27, 1879, first—two males; both sexes common March 31. Rarely seen in the fir forests, very common in valleys and foothills in summer; common on borders of Sierra Valley, June, 1885.

Murphy's. J. P. Snyder.—April 5, two; common April 16, 1885. Last seen August 27; rare on the 12th. First seen 1886, April 3.

Sebastopol. F. H. Holmes.—Common summer resident. First seen April 22, 1885.

Marysville. W. F. Peacock.—April 6, 1884, first male; bulk arrived April 20. In 1885, first seen March 27; common April 4; common summer resident.

Gridley. L. B.—April 3, 1890, four males together.

Chico. Wm. Proud.—April 5, 1884, first. In 1885, first seen March 25. No doubt an old friend because of his serenading us from the weeping willow at the end of the house, where, for nine consecutive years, a brood has been raised and some seasons two broods.

Willamette Valley. O. B. Johnson.—Summer, common; breeding extensively.

Cooper, 1860. Does not arrive at Puget Sound until the beginning of June, and is not very common there.

British Columbia. John Fannin.—Summer resident; not common.

Henshaw, 1879. Very numerous about Carson and to the northward along the base of the mountains to the Columbia River.

Dalles. Suckley, 1860.—May 7, male specimen.

Camp Harney. Bendire.—Common during the summer months; arrives about May 10.

Ridgway. Common in all the wooded localities of the western country.

Cooper, 1870. Fort Mojave, I saw none until April 1, 1861. Santa Cruz, April 3, 1866.

141. **Scolecophagus cyanocephalus** (Wagl.) Brewer's Blackbird.

Cape region, winter, at La Paz and San Jose del Cabo, a large flock in the streets and yards at the latter locality, associating as at Guaymas with the dwarf cowbird; this flock scattered in pairs in the willows along the river in May. San Rafael, May 17, tolerably common; also south of Campo toward Hansen's in suitable localities. San Diego, common resident.—L. B.

Poway. F. E. Blaisdell.—Common summer resident.

Poway. W. O. Emerson.—April 7, nesting. Volcan Mountains, three males came around the house in the snow storm of February 11, and a female was seen March 20.

San Bernardino. F. Stephens.—Common summer resident of the valley; less common in winter in valley and foothills.

Agua Caliente. F. Stephens.—Several seen March 21, 30, April 10 and 11, 1886.

Henshaw, 1876. Very abundant and constant resident. Many were breeding in company with a large colony of the *A. tricolor* before mentioned.

Santa Cruz. Joseph Skirm.—Very common.

Alameda and Contra Costa counties. W. E. Bryant.—Common resident.

Berkeley. T. S. Palmer.—Common summer resident. First, May 11, 1886.

Central California. L. B.—Very common in summer excepting in the mountains and there usually found about meadows of any considerable extent. Very abundant in the valleys in winter, breeding plentifully in towns as well as in the country, nesting early in trees of different kinds, especially cultivated evergreens.

Willamette Valley. O. B. Johnson. Very abundant in summer; breeding abundantly.

Cooper, 1860. They are found throughout the Territory. Vancouver, winter.

Suckley, 1860. Quite abundant at Fort Dalles; a winter resident.

British Columbia. John Fannin.—Rare summer resident.

Henshaw, 1879. Extremely abundant throughout this whole region as a summer visitant, while more or less remain through the winter.

Camp Harney. Bendire.—An exceedingly abundant summer resident, a few remaining during mild winters. It breeds in various situations on the ground, in sage bushes and in service-berry bushes.

Ridgway. Seldom seen there during summer; this blackbird becomes one of the most abundant species in the lower valleys during the winter season, when immense flocks frequent the settlements.

Hoffman. Met with more particularly in the southwestern portion of Nevada, in the more elevated regions.

142. Coccothraustes vespertina (Coop.) EVENING GROSBEAK.

British Columbia. John Fannin.—Breeds east of the Cascades; accidental west, though sometimes found on Vancouver's Island.

Dr. Cooper, 1860. Common resident of the forests. January, 1854, a flock at Vancouver.

Walla Walla. J. W. Williams.—April 5-7, 1885, about fifty; April 10, next and last seen. It is rare here; does not breed. These passed on north with the thistle-bird *(S. tristis)*.

O. B. Johnson, 1880. Sometimes plentiful during the spring migrations.

L. B.—Sierra City to Sierraville, June 18, 1885, a flock. Summit, Central Pacific Railroad, July 8, a large flock. Blood's, Big Tree and Carson road, altitude 7,200 feet, rare but regular summer resident. Mr. Blood informed me in the summer of 1879 that several females reappeared around his corral and dwelling on July 22 or 23 with young. I was there July 16 and saw four adult males but no females. In July, 1880, after much searching for the nest of the only pair I found here this season, I concluded that it was inaccessible in a large fir tree. Dr. A. C. Davenport received in March eight specimens in flesh collected near Murphys by Mr. Thomas Goodwin, an old resident, to whom they were novelties. I have seen it in summer at several localities in Calaveras and Alpine counties, but it is rare so far south.

Murphys. John J. Snyder.—October 8, 1885, first evening grosbeak. They became common soon after.

Santa Cruz. A. M. Ingersoll.—November 5, 1885, eight or ten seen, and Mr. George Ready tells me he saw a large flock on San Lorenzo River, November 1.

Sebastopol F. H. Holmes.—I shot two December 7, 1885.

143. Pinicola enucleator (Linn.) PINE GROSBEAK.

British Columbia. John Fannin.—Big Bend of Columbia River, rare.

Camp Harney. Bendire.—An occasional winter visitor.

L. B.—Summit, Central Pacific Railroad, August 11, 1882, tolerably common; from June 23 to July 10, 1885, an adult male and female feeding in alders; during this time these only; but later, in August and September, not rare, in fact rather common. Blood's, July 16, 1880, shot an adult female which probably had a nest; specimen sent to Smithsonian Institution.

144. Carpodacus purpureus californicus Baird. CALIFORNIA PURPLE FINCH.

San Diego, January 19, 1884, shot a female; not met here again, nor at Campo in January. Mr. Emerson did not find it or *C. cassini* on the Volcan Mountains in the severe winter of 1883-84. It is common in Central California from 3,000 feet altitude up to about 5,000 feet in summer, and much lower in winter, rarely visiting the large interior valleys near sea level at that time. L. B.

San Bernardino. F. Stephens.—Rare accidental visitant to the foothills. Agua Caliente, seen a number of times from March 18 to April 12, 1886.

Santa Cruz. Joseph Skirm.—Tolerably common summer resident.

Berkeley. T. S. Palmer.—Rare. December 26, 1885, three birds.

Oakland and vicinity. W. E. Bryant.—Rare summer resident.

Olema. A. M. Ingersoll.—Breeds.

Beaverton. A. W. Anthony.—Abundant summer resident; March 6, first; March 20, 1884, common;

April 1, bulk arrived; February 20, 1885, ten birds; next seen February 23; common March 27. Very common; breeds.

Seattle. O. B. Johnson.—March 14, 1884.

Cooper, 1860. Abundant. A few remain all the year in the Territory.

British Columbia. John Fannin.—Common summer resident.

Henshaw, 1879. A single individual taken at the Dalles in October, the first record at any point along the eastern slope.

Burrard Inlet. John Fannin.—April 29, 1885, first; May 10, next; May 16, common. Very common in breeding season.

Willamette Valley. O. B. Johnson.—Common summer resident.

145. Carpodacus cassini Baird. CASSIN'S PURPLE FINCH.

Calaveras, Alpine, Placer, Nevada, Sierra and Butte counties, common in breeding season from 4,700 feet altitude upward; not collected on the west slope in winter nor reported from Mt. Whitney by Mr. Henshaw in autumn, by Mr. Stephens in San Bernardino Mountains at any time, nor by Mr. Emerson in the Volcan Mountains, altitude 6,000 feet, in the severe winter of 1883–4. Xantus found it at Fort Tejon. I found a flock of about a hundred all in female plumage, though some were males, in mountains near Tehachapi, March 29, 1889.—L. B.

Ridgway. Near Carson, first observed March 21; they continued to increase in abundance until about the middle of April; eastern slope of Ruby Mountains, quite abundant on several occasions.

Henshaw, 1879. Abundant summer inhabitant in the

neighborhood of Carson, becoming rather less numerous as northern California is reached, the species persistent to the Columbia River.

Camp Harney. Bendire.—Moderately abundant summer visitor; breeds in the pine forests of the Blue Mountains. It is probable that some remain throughout mild winters.

British Columbia. John Fannin.—Common summer resident.

146. Carpodacus mexicanus frontalis (Say). HOUSE FINCH.

San Diego. Abundant resident; young out of the nest April 24; young common May 1.—L. B.

Poway. F. E. Blaisdell.—Common resident.

Volcan Mountains. W. O. Emerson.—January 28, February 22 warm and sunny, celebrating the day with song.

San Bernardino. F. Stephens.—Abundant resident of the valley. Common resident of the foothills. Agua Caliente, abundant; probably resident. March 18 to April 15, 1886, common resident.

Henshaw, 1876. Perhaps the most numerous of any of the small birds.

Alameda and Contra Costa counties. W. E. Bryant.—Abundant resident.

Berkeley. T. S. Palmer—Abundant resident; nest and five fresh eggs April 11, 1885; incubation lasts eleven days; young fly from 16 to 18 days after hatching. Abundant about March 21, 1886.

Sebastopol. F. H. Holmes.—First seen February 22, 1885; next February 25; common March 1. Breeds abundantly.

Ukiah. G. E. Aull.—Abundant summer resident.

British Columbia. John Fannin.—Summer resident; not common.

L. B.—No doubt abundant in all the principal agricultural districts of California; winters at Murphys, Colfax and head of Sacramento Valley, not breeding in the Sierra on the west slope much above 3,000 feet, but much higher in summer on the east slope. In ascending the west slope of the Sierra Nevada in Central California in summer, the three species are found occupying separate belts, *C. m. frontalis* the lower, *C. purpureus californicus* intermediate, and *C. cassini* the upper belt, but meeting *californicus* on its lower range; the lower species being the smallest, the higher the largest.

Henshaw, 1879. Not observed much farther north than Honey Lake, northern California. An abundant species wherever found.

Camp Harney. Bendire.—April 8, 1876, a single specimen.

Ridgway. Carson, May 13, 1868, first seen. Although chiefly a bird of the lower valleys, sometimes found in the lower cañons of the mountains, Sacramento, Truckee Valley, Pyramid Lake, West Humboldt Range, etc.

Hoffman. Rather common and generally distributed over the northern half of the region under consideration; Spring Mountain near the Old Spanish Trail.

147. Loxia curvirostra minor (Brehm). AMERICAN CROSSBILL.

British Columbia. John Fannin.—Common summer resident.

Cooper, 1860. Abundant in winter near the coast.

O. B. Johnson. Common in the mountains and coming down to the valley in winter.

Summit, Central Pacific Railroad. L. B.—November 12, 16, 1884, three flocks; same locality, September 9, 1885, first, two flocks; the species tolerably common afterward until I left, October 12. I think a few breed in Calaveras County, as I have seen a very few in the fir forests in breeding time.

Ridgway. None seen until towards the last of August when they gradually became common in the Humboldt Mountains. August 12 a male, *leucoptera* probably, seen on the east slope of the Ruby Mountains.

Newberry. A constant feature of the pine forests of Oregon and northern California.

Henshaw, 1879. Becomes numerous in the mountains and on the foothills in fall.

Camp Harney. Bendire.—Common during the winter months in large flocks.

148. **Leucosticte tephrocotis** Swains. GRAY-CROWNED LEUCOSTICTE.

Camp Harney. Bendire.—A winter visitor, associated with *L. littoralis* from November 8, 1875, to March 22, 1876. Specimens at different times.

149. **Leucosticte tephrocotis littoralis** (Baird). HEPBURN'S LEUCOSTICTE.

Burrard Inlet. John Fannin.—I shot two here November 14. This is the first time I ever saw these birds at sea level. Common summer resident in British Columbia east of the Cascades.

Camp Harney. Bendire (Birds Southeastern Oregon).—I have observed these birds almost daily for two winters, and examined about two hundred specimens. It is probable that some breed on high peaks.

Virginia City. Ridgway.—A single flock of this species was seen on the 5th of January. The flock comprised about fifty individuals.

150. **Acanthis linaria** (Linn.) REDPOLL.

British Columbia. John Fannin.—Breed north; remain here late in fall, etc.

Dr. Cooper, 1860. Washington Territory.—A small flock appeared on the coast in winter. I obtained one specimen. They fed on alder and thistle seeds.

Camp Harney. Bendire.—Found in large flocks during winter.

Fort Klamath. Lieutenant Wittich (Bull. Nutt. Orn. Club, iv).—May 9, 1878.

151. Spinus tristis (Linn.) AMERICAN GOLDFINCH.

San Diego. Rare, winter of 1883–84.

Tia Juana, N. W. Lower California, April 30, 1885, large flock.—L. B.

Poway. F. E. Blaisdell.—Common summer resident. Seen February 17, 1884.

San Bernardino. F. Stephens. — Tolerably common summer resident of the valley and foothills.

Oakland. W. E. Bryant.—Tolerably common summer resident. First seen in 1886, on April 7.

Berkeley. T. S. Palmer.—Tolerably common summer resident.

Central California. L. B.—Tolerably common resident in valleys and foothills. Red Bluff, February 3, 1885.

Willamette Valley. O. B. Johnson.—Common summer resident.

Beaverton. A. W. Anthony.—February 20, 1884, ten seen; next seen February 23; common March 27; very common in breeding season.

Walla Walla. J. W. Williams. — March 27, 1885, males; April 5, males; May 26, one, the last.

Burrard Inlet. John Fannin.—April 4, 1885, first seen; next April 7; common April 10. Tolerably common summer resident of British Columbia.

Henshaw, 1879. Common summer resident at many points. Found on the Columbia River in October.

Camp Harney. Bendire.—Seen on but a single occasion, May 5, 1876.

Ridgway. Eastward of the Sierra Nevada extremely rare. Truckee Valley, breeding; rare east slope Ruby Mountains.

Smoky Valley. Hoffman.—Less than half a dozen individuals.

Beaverton. A. W. Anthony.—April 18, 1885, first; common May 5; breeds. Very common.

152. Spinus psaltria (Say). ARKANSAS GOLDFINCH.

The species is rare in the northern 100 miles of Lower California in May, most numerous in the mountains, and this I think applies to San Diego County below about 4,000 feet altitude, where not rare in winter.—L. B.

San Bernardino. F. Stephens.—Tolerably common resident of the valley and foothills. Agua Caliente, common March 25–28; March 18 to April 15, 1886, common resident.

Henshaw, 1876. Of the three species inhabiting southern California this goldfinch appears to be the most widely spread and perhaps the most numerous; like the other two it inhabits the valleys.

Santa Cruz. Joseph Skirm.—Common summer resident.

Berkeley. T. S. Palmer.—Abundant resident.

Alameda. A. M. Ingersoll.—January 7, 1885, common.

Central California. L. B.—Common resident of foothills, less common in the valleys, rare in the pine forests where perhaps it is only found when migrating. At Summit, altitude 7,000 feet, August 27, 1885, I saw a large flock in which were a dozen or more *S. pinus*, and these I thought were crossing from the east to the west slope; also noticed in small flocks in September. Red Bluff, February 3.

Siskiyou County, Cal. Baird, Brewer and Ridgway.

153. Spinus lawrencei (Cass.) LAWRENCE'S GOLDFINCH.

San Diego, common resident; March 22, making nest in a small orange tree. Colonel N. S. Goss noticed it nesting here twelve days earlier.—L. B.

Poway. F. E. Blaisdell.—Common summer resident; May 17, several nests in which the young were nearly fledged.

Volcan Mountains. W. O. Emerson.—January 25 and 31, a small flock; April 5, common at Live Oak Springs.

San Bernardino. F. Stephens.—Tolerably common summer resident of the valley and foothills. Agua Caliente, March 25–28; April 6, 12, 14 and 15, 1886.

Henshaw, 1876. Near Santa Barbara, the only place where I met with this bird, it was a numerously represented species.

Santa Cruz. Joseph Skirm.—Common summer resident.

Alameda and Contra Costa counties. W. E. Bryant. Rare summer resident.

Berkeley. T. S. Palmer.—Usually rare summer resident. Common in 1885.

Nicasio. C. A. Allen.—May 10, 1884, first.

Sebastopol. F. H. Holmes.—Tolerably common summer resident. May 15, 1885, first seen; seen as late as October 25.

Marysville. L. B.—January 9, 1878, rare, in freezing weather.

Chico. Wm. Proud.—May 22, 1884, nest and young.

Heermann. Very abundant throughout the northern mining regions of California.

Cooper, 1870. I found a few of the species at Fort Mojave.

154. Spinus pinus (Wils.) PINE SISKIN.

British Columbia. John Fannin.

Cooper, 1860. An abundant and constant resident of Washington Territory.

Seattle. O. B. Johnson, 1884.—May 1, mating and nesting

Walla Walla. J. W. Williams.—March 26, 1885, first, six birds; next seen March 27, April 3.

Beaverton. A. W. Anthony.—March 14, 1885, first; March 15 common; last seen April 30.

Willamette Valley. O. B. Johnson, 1880. Common winter resident.

Central California. L. B.—Rare summer resident of the high Sierra from Calaveras County north; rare at Butte Creek House July 1-2. Common during migrations in the pine region; common in winter in the foothills, associating much with *S. psaltria* and perhaps hybridizing with it occasionally. I have seen two which I thought were crosses, one of them having been shot at Murphy's and sent to the Smithsonian in 1878. Seldom seen in the valleys in winter. A single specimen shot at La Paz, in Lower California, in winter of 1882.

Sebastopol. F. H. Holmes.—Abundant winter visitant. September 3, first; April 5, last.

Berkeley. T. S. Palmer.—Common March 7, 1885; rare winter visitant. A flock of six were seen January 9, 1886.

Oakland and vicinity. W. E. Bryant.—Common winter visitant.

Santa Cruz. Joseph Skirm.—Common.

Kernville. Henshaw, 1876. Last of October present in small flocks.

San Bernardino. F. Stephens.—Rare winter visitant to mountain. Agua Caliente, flock seen April 12, 1886.

Ridgway. In summer abundant in all the pine forests from the Sierra Nevada to the Uintah.

Henshaw, 1879. Occurs along the whole eastern slope in fall and winter and passes the summer in the coniferous belt in much of its extent.

Camp Harney. Bendire.—Common during the winter months in the Blue Mountains.

Hoffman. Occur during the summer in all the pine forests.

San Jose. W. O. Emerson.—A large flock, September 17, 1886.

155. Plectrophenax nivalis (Linn.) SNOWFLAKE.

British Columbia. John Fannin.—Resident east of the Cascades.

Camp Harney. Bendire.—Found sparingly during the winter months.

156. Calcarius lapponicus (Linn.) LAPLAND LONGSPUR.

Camp Harney. Bendire.—Found sparingly during the winter months.

Carson. Ridgway.—Frequently detected among the large flocks of horned larks during the more severe parts of winter.

157. Poocætes gramineus confinis Baird. WESTERN VESPER SPARROW.

San Diego. L. B.—Tolerably common in winter in valleys back from the coast.

Poway. F. E. Blaisdell.—Common in January and February.

San Bernardino. F. Stephens.—Rare transient visitant to the valley.

Newberry. Common in the Sacramento Valley. [Winter?]

L. B.—Probably found in the lower portions of Central California in winter only, and during its migrations; even then rare. At North American Hotel, thirty miles east of Stockton, I found a flock of forty or fifty apparently wintering as they were there nearly all of January, 1885. At Gridley I found a few November 10, and again December 11, 1884; have seen a few in fall in the subalpine valleys of Calaveras, Alpine and Placer counties, and found it breeding and common in Sierra Valley, which, however, has east slope characteristics and some of its species, including an abundance of sage brush and sage hens. First seen at Summit during the fall migration, September 9, 1885.

Sebastopol. F. H. Holmes.—September 27, 1884, first; rare winter visitant.

Willamette Valley. O. B. Johnson.—Common during the summer.

Beaverton. A. W. Anthony.—Common summer resident; April 5, 1884, May 1, abundant.

Washington Territory. Cooper.—Common in summer on the prairies of the interior.

British Columbia. John Fannin.—Transient visitant.

Henshaw 1879. Numerous in the valleys; noticed no farther north than southern Oregon.

Camp Harney. Bendire.—Very common summer resident, breeding abundantly.

Ridgway. Most frequently met with during summer on the open grassy slopes of the higher cañons in September, becomes exceedingly abundant along the foothills of the higher ranges; appears to make a complete southward migration none having been seen at Carson until the first of April.

Cooper, 1870. I found this bird wintering in the Colorado Valley in considerable numbers but it disappeared by April 1.

158. Poocætes gramineus affinis Miller. OREGON VESPER SPARROW.

In the winter of 1883-4 I got a specimen in Cajon Valley, San Diego County, out of a flock of larger grayer birds and sent it to Mr. Ridgway, who said that it was probably deserving of a name but he would want to see more of that sort before naming it, and I let the matter rest until the fall of 1888 when I sent several to Dr. Fisher who was absent from his post about that time. In the meantime Mr. Miller's bird was published. Dr. Fisher told me afterward that he compared my specimens with Mr. Miller's and found them identical, consequently the preceding notes on *P. g. confinis* belong in part here. I collected the one I sent to the Smithsonian in 1884; selected it from a flock of about a dozen, because it was smaller and differed decidedly in color from its companions.

The birds which I have shot in winter in Central California are all, or nearly all, of the variety *affinis*.— L. B., April, 1890.

159. Ammodramus sandwichensis (Gmel.) SANDWICH SPARROW.

British Columbia. John Fannin.—Common summer resident.

Cape Beale, B. C. Emanuel Cox.—Very common all the year around; the only birds that strike the light. Specimens identified by Mr. Ridgway were typical *sandwichensis*.

Willamette Valley. O. B. Johnson.—Seen sparingly during the migration.

Butte County, California. L. B.—Winter. Specimens collected by me very near identical with an Alaskan specimen kindly furnished by Mr. Henshaw, having bills of equal size, and equally as large in all respects.

160. Ammodramus sandwichensis alaudinus (Bonap.)
WESTERN SAVANNA SPARROW.

The form which I consider typical *alaudinus* was rare in mid-winter at San Diego, though it goes as far south as La Paz and Cape St. Lucas in winter. In March it became more common, and was so until April 20. April 23 last seen, all having gone north of San Diego to breed. In winter it is abundant in many parts of California, breeding on the east slope of the Sierra, in latitude 39° and northward.—L. B.

Poway. F. E. Blaisdell.—Common in January and February.

Volcan Mountains. W. O. Emerson.—March 9, first. A specimen taken on the 11th was much smaller and grayer than that of the 9th. Some days after this date it was in large flocks in Santa Isabel Valley.

San Bernardino. F. Stephens.—Rare winter visitant. Agua Caliente. Not Common.

Henshaw, 1876. Early in September it was high up in the mountains near Mount Whitney, while in November it was exceedingly numerous about Oakland.

Oakland and vicinity. W. E. Bryant.— Common winter visitant.

Sebastopol. F. H. Holmes.—Found breeding abundantly, June 10, at Sebastopol. *(bryanti?)*

Santa Cruz. Joseph Skirm.—Summer resident. *(bryanti?)*

Henshaw, 1879. Numerous in all wet meadow lands, and as much so in Oregon as farther south. A nest taken near Washoe Lake, May 22, contained four fresh eggs.

Camp Harney. Bendire.—Very common in the early spring.

Ridgway. Abundant in every moist meadow and grassy marsh, from March to November inclusive.

161. Ammodramus sandwichensis bryanti Ridgw.
BRYANT'S MARSH SPARROW.

Oakland and vicinity. W. E. Bryant.—Common resident. L. B. Probably some of the notes under *A. s. alaudinus* belong here, and possibly some of those under *A. sandwichensis* belong to *A. s. alaudinus*.

Mr. F. G. Holmes sent me a juvenile specimen which was reared at Sebastopol, and which I think belongs to *A. s. bryanti*.

162. Ammodramus beldingi Ridgw. BELDING'S MARSH SPARROW.

San Quintin Bay, 150 miles south of San Diego, May, 1881, very common and the only one of the genus found by me in May, 1881.

Todos Santos Islands. Dr. Streets (Bull. 7, Nat. Mus.) one specimen. These islands are about sixty miles southward from San Diego, near the Pacific Coast of Lower California.—L. B.

San Diego Bay. L. B.—Very common resident.

Santa Barbara. Henshaw, 1876. June and July, 13 specimens, in worn plumage.

Port Harford. L. B.—May 20, 1884. A single male which probably had a mate nesting. As in May, 1885, I could not find one of the species or genus here, I suppose this to be the northern limit of its range, while San Quintin Bay is the most southern point known. It was mated at San Diego, March 25, 1885, where, April 4, I found a nest and two eggs, bird sitting four or five days. The first young out of nest May 1st. No song traced by me to this bird, but Mr. Holterhoff, who for months has nearly every morning and evening passed through the marshes between San Diego and National City, says he has often heard them sing. This appears to be, the only *Ammodramus* which breeds in San Diego County. The nest found April 4 was near the beach just above high tide in a dense growth of Salicornia, Atriplex and Frankenia. It was composed of surf-worn eel grass mostly, lined with a few feathers. The eggs are of a faint greenish, or bluish white ground color, one being paler than the other and perhaps unfertile, both heavily blotched with brown of nearly uniform shade, confluent and hiding the ground color at the large end. Measurement, 80 x 55—80 x 56. It frequently runs on the sandy beach, not far from cover, when pursuing insects, quite as rapidly as a sandpiper.

Mr. Henshaw's Santa Barbara specimens being in worn and faded plumage, as the species usually is in June and July, did not attract his attention as they must otherwise have done.

The first specimens I collected were taken at San Quintin Bay, where no other *Ammodrami* were present to confuse me. One of these I labeled *P. guttatus*, with an interrogation. Afterward, having given it much attention about San Diego in the winter of 1883–4, I became convinced that it was a distinct species.

163. **Ammodramus rostratus** Cass. LARGE-BILLED SPARROW.

Common in winter in the Cape Region, at San Diego and San Pedro during the same time, and probably formerly bred at the two latter localities. I could not find the species about San Diego Bay or False Bay in April and May, 1881, nor in April and May of the years 1884 and 1885, in the latter year having followed the coast nearly fifty miles north of San Diego without finding it. I last saw it at San Diego, March 10, 1884. Its nesting places and nesting habits are still unknown, although Dr. Cooper found them feeding their young at San Pedro in July but never found a nest that he was certain belonged to this species (Cal. Orn., 1870).

I have noticed the species with more than ordinary interest at Guaymas, La Paz, San Jose del Cabo, Todos Santos Village, San Diego and San Pedro, and having seen the type of *guttatus*, think *rostratus* a species of but little individual variation, even including *guttatus* and Dr. Streets' San Benito Island specimens. It is a species that can hardly be traced to its origin.

164. Ammodramus bairdii (Aud.) BAIRD'S SPARROW.

Camp Harney. Bendire.—May 24, 1876, I took a nest and eggs with the parent, which I identified as this species.

165. Ammodramus savannarum perpallidus Ridgw. WESTERN GRASSHOPPER SPARROW.

San Diego, April 19, a male, first; probably a few breed in El Cajon.—L. B.

Santa Barbara. Henshaw, 1876.—Directly on the coast, breeding.

Ridgway. From June 6 to July 4, 1867. Abundant in the fields about Sacramento City, as well as throughout the interior.

Oakland. W. E. Bryant.—Five male specimens in spring.

Hoffman. Found to breed near Eureka, Nevada.

166. Chondestes grammacus strigatus (Swains.) WESTERN LARK SPARROW.

San Diego. Common resident.—L. B.

Poway. F. E. Blaisdell.—Common resident.

Volcan Mountains. W. O. Emerson.—Altitude 6,000 feet, February 24, large flocks feeding with shore larks on open flats; seen off and on during my stay on the Volcan; common all through Santa Isabel and Santa Maria valleys.

San Bernardino. F. Stephens.—Common resident of the valley.

Santa Cruz. Joseph Skirm.—Common resident.

Alameda City. A. M. Ingersoll.—Several seen December 31.

Alameda and Contra Costa counties. W. E. Bryant.—Tolerably common resident.

Berkeley. T. S. Palmer.—Summer resident. First seen March 22; next seen March 25; common April 20, 1886.

Sebastopol. F. H. Holmes.—Abundant summer resident. March 4, 1885, first; March 15, common. I think it does not winter here.

Ukiah. G. E. Aull.—Abundant resident.

Central California. L. B.—Common constant resident of the valleys and foothills, including Murphys and Colfax, both about 2,500 feet altitude, and rather common at Red Bluff, February 3-5, 1885. Perhaps our birds go south in winter and those from Oregon take their places, but the species is here at all times.

Willamette Valley. O. B. Johnson.—Sparingly common during the summer.

Suckley, 1860. One specimen at Fort Dalles; not seen west of the Cascades.

British Columbia. John Fannin.—Summer resident; not common.

Henshaw, 1879. Lost sight of about 100 miles north of Carson.

Bendire. In the immediate vicinity of Camp Harney few specimens; at Juniper Lake, 80 miles south, quite a number breeding.

Ridgway. Exceedingly abundant in favorable localities throughout the entire extent of the western region. Specimens at Sacramento, Truckee and Salt Lake.

Hoffman. Common in favorable localities over nearly the whole of Nevada.

Walla Walla. J. W. Williams.—June 2, 1885, two; next seen June 10; both specimens shot; not known if it breeds.

167. Zonotrichia leucophrys (Forst.) WHITE-CROWNED SPARROW.

San Diego, May 3, 1885, male, first; May 5, female, ovaries very small; only about a dozen of the species seen this spring, none at any other time, neither in spring of 1881 nor 1884. None seen by the Goss Brothers at Julian from March 17 to May 15, 1884.—L. B.

Poway. F. E. Blaisdell.—October 3, 1884, first; October 6 a flock [the first southern California fall record].

Henshaw, 1876. I found it in the high Sierra (Mt. Whitney) in September in company with *Z. intermedia*, forming, however, but a very small proportion of the vast flocks of those birds.

Blood's, Alpine County, altitude 7,200 feet. W. E. Bryant.—Common, June 26, 1885, breeding.

L. B.—Common in summer in the subalpine meadows from Alpine County to the northern part of Butte County; probably has a much more extensive breeding range in the Sierra than is here indicated; appears to winter entirely south of California; quite common in Amador and Sierra counties on the east slope in breeding season. Summit Meadows, Donner Pass, June 24, 1885, a pair beginning nest; July 7, nest and nearly fresh eggs, both on the ground, but I have found nests in willows. Blood's, July 9, 1880, still unmated; willows still destitute of leaves; a few snow banks in the meadow, which, however, is in many places yellow with buttercups *(Ranunculus)*. Upon returning, August 10, I could not find young birds, though in the preceding year on July 16 they were large enough to tumble out of a nest as I approached it. This shows the difference in seasons in these high mountains consequent upon difference in snowfall. When the weather does become favorable vegetation grows with an astonishing rapidity in the long days of June and July, hardly waiting for the snow to melt; in fact, sometimes bursting through it, and about the middle of July, 1880, I was agreeably surprised to find on the top of the mountain about two miles north of Blood's, at a height of nearly 9,000 feet, a patch of an acre or two of the seemingly delicate *Claytonia carolinensis* (var.?) in perfect flower standing in compact snow three or four inches deep, a part of the previous winter's product. Where the snow was so deep that it covered them they were also flowering, with a vacant cylindrical space about an inch in diameter immediately around them and a thin, icy, bubble-like cover on the surface; in reality a miniature hot-house.

Henshaw, 1879. As almost everywhere throughout the west, this sparrow occurs along the east slope in great numbers during the migrations. It is also numerous in these mountains in summer.

Fort Klamath. Wittich, Nutt. Bull., iv, 165. Numerous; specimens April 26, 1875, April 29, 1878.

168. Zonotrichia leucophrys intermedia Ridgw. INTERMEDIATE SPARROW.

At San Diego it is an abundant winter visitant; last seen here April 23, 1884; was common to the 15th.—L. B.

Poway. F. E. Blaisdell.—Common winter visitant; not seen since the last of April.

Volcan Mountains. W. O. Emerson.—January 24, 1884, to February 14, in large flocks, when the snow drove them down to the lower cañons and valleys. Afterward a few were seen when the sun shone and it was warm. It was observed at Poway as late as April 27.

Mt. Whitney. Henshaw, 1876.—Middle of September large flocks.

L. B.—This form has been very abundant in winter wherever I have been in California, if below the snow line. It probably breeds entirely north of California. Probably some of the notes on *Z. l. gambeli* really refer to this form, which undoubtedly inhabits all of the low parts of California in winter vastly outnumbering all the other *Zonotrichiæ*. It was not separated from *gambeli* until 1873.

Oakland and vicinity. W. E. Bryant.—Winter visitant.

Henshaw, 1879. Flocks frequently fall into the path of the collector in this region (east slope) during spring and fall. Not found in the mountains as a summer resident, and I believe it goes farther north to breed.

Camp Harney. Bendire.—A moderately abundant summer resident.

Ridgway. West Humboldt Mountains, September 7, 1867. Truckee, December 26. Head of Humboldt Valley, September 16, 1868; abundant at Carson in willows and brushwood in winter.

Hoffman. Met with several times in more southern

portions of Nevada, particularly in the elevated and fertile valleys just north of Mt. Magruder; again sparingly east of Spring Mountain in September.

Haywards. W O. Emerson.—September 24, 1884, first; common at once.

Berkeley. T. S. Palmer.—A pair secured April 13, 1886. I do not think this species has ever been taken here before.

Poway. F. E. Blaisdell.—October 23, 1884, first; common all winter.

L. B.—Summit; September 12, 1885, three specimens; common at once; rare October 11. The first arrivals at Summit appeared to be juveniles. Big Trees, September 25, 1880, first. Stockton, September 18, 1881, first; rare. Stockton, September 20, 1883, several. Gridley, September 24, 1884, first; common September 25.

Walla Walla. D. T. Williams.—April 3, 1885, eight specimens; not very common summer resident.

169. **Zonotrichia leucophrys gambeli** (Nutt.) Gambel's Sparrow.

Julian. N. S. Goss.—April 9, 1884, I saw three *Z. gambeli*, the only ones I saw from April 1 to May 16; was in the field every day.

San Bernardino. F. Stephens.—This species with var. *intermedia* are abundant winter visitants to the valley and foothills. Agua Caliente. Abundant March 25–28, 1884. Abundant winter resident. Common March 18 to April 15, 1886.

Mt. Whitney. Henshaw, 1876.—September; only two specimens.

Santa Cruz. Joseph Skirm.—Common; stays the whole year in the low brush along the beach; begins to breed about April 25; eggs from three to four; raises two, possibly three broods in a season.

Alameda. A. M. Ingersoll.—July 2, 1884, I found one nest with three fresh eggs; one just finished without eggs, the latter within a few feet of one I found March 29. This species must raise two or three broods here in summer.

Oakland and vicinity. W. E. Bryant.—Common resident.

Nicasio. C. A. Allen.

Sebastopol. F. H. Holmes.

Butteville, Or. W. E. Bryant.—In breeding season of 1883.

Willamette Valley. O. B. Johnson.—A very common summer resident and nesting familiarly about gardens and thickets near dwellings.

Beaverton. A. W. Anthony.—Common summer resident. April 7, first; April 12, bulk arrived. May 12, first nest. First seen March 25, 5 birds; common March 30, 1885.

Cooper, 1860. Abundant in all the prairie districts; frequents the coast prairies where I have found its nest and eggs.

Suckley, 1860. Very abundant at Fort Dalles and Puget Sound, a constant summer resident at both places.

British Columbia. John Fannin.—" var. *gambeli*," summer resident east of the Cascades. Arrived at Burrard Inlet April 22, 1885. Only a few noticed and found breeding here for the first time.

L. B.—I have never found this form in central California but once, and then only a few individuals at Stockton in spring, migrating. At Port Harford, May 26, 1885, I found about a dozen near the beach. These were mated and undoubtedly breeding. At this place in May, 1884, I shot one which I was inclined to refer to *intermedia*, but I suppose the coast breeding birds on the California and Oregon coasts may be considered as the typical form, though I have seen but few specimens from the coast. One of Mr. Anthony's Beaverton specimens was nearly typical *gambeli*, and I supposed the specimen represented the birds which breeds there. I could not find it about San Diego.

170. **Zonotrichia coronata** (Pall.) GOLDEN-CROWNED SPARROW.

San Diego. Rare winter visitant; last seen April 3, 1884; April 17, 1885. Probably goes but little south of San Diego.—L. B.

Poway. F. E. Blaisdell.—Rare winter visitant; not noticed after the last of March.

Volcan Mountains. W. O. Emerson.—Common from January 24, 1884, to February 14, when the snow appeared to be too much for them. A few seen in February and March after the storm of February 14, especially during warm days.

San Bernardino. F. Stephens.—Very rare winter visitant to the foothills.

San Jose. A. L. Parkhurst.—May 4, 1884, last.

Alameda and Contra Costa counties. W. E. Bryant. Tolerably common winter visitant.

Berkeley. T. S. Palmer.—Abundant winter visitant.

Central California. L. B.—Abundant winter visitant below the snow line. Big Trees, September 25, 1880, first; common at Murphys on the 29th. Gridley, September 24, 1884, first, six seen; common on the 25th. Summit, September 25, 1885, first, rare; common October 2 (snow on September 24). It left Stockton after April 27, 1880—a late spring.

Sebastopol. F. H. Holmes.—Abundant winter visitant; September 24, first; last seen May 2, bulk departed about April 15, 1884.

Shasta County, Cal. Brewer.—June 14, 1877, nest

and eggs. (Bull. N. O. C., Vol. 3, 42.) [This is the only reliable record of its breeding in California.]

Willamette Valley. O. B. Johnson.—Sparingly common during summer.

Beaverton. A. W. Anthony.—One specimen as late as May 22, 1884, but mostly gone by May 1.

Cape Flattery Light. Alexander Sampson, keeper. April 26, 1884, a few sparrows with black on the sides of the head and a yellow stripe on top. May 6, quite a flight of the same birds; ten killed by striking the light. Wind, southeast, 38 miles an hour.

Suckley, 1860. Not rare in the vicinity of Fort Dalles or Fort Steilacoom; in both places quite abundant in summer.

Cooper, 1860. I saw them but once near Puget Sound, on the 10th of May.

British Columbia. John Fannin.—Rare summer resident. Burrard Inlet, September 13, arrived; September 29, here in numbers. I see it here only in fall.

Henshaw, 1879. Occurs along the eastern slope during the fall migration. Its numbers are limited to the comparatively few that find their way into the bands of white-crowned and Ridgway's sparrows [*intermedia*].

West Humboldt Mountains. Ridgway.—October 7, one specimen; the only one seen.

171. Zonotrichia albicollis (Gmel.) WHITE-THROATED SPARROW.

British Columbia. John Fannin.—Rare summer resident.

Henshaw, 1879. About forty miles from the Dalles, one of two seen, shot. I presume that it is by no means scarce.

172. **Spizella monticola ochracea** Brewst. WESTERN TREE SPARROW.

British Columbia. John Fannin.—Common summer resident.

Fort Walla Walla. Winter of 1881-82. Bendire.—(Bull. N. O. C., vol. 7, 225.)

Henshaw, 1879. Seen upon the Columbia River in October.

Camp Harney. Bendire.—Moderately abundant during the winter months.

Hoffman. Common throughout the northern regions; farther south more sparingly and only in the more elevated and fertile areas.

Ridgway. During the winter common and very generally distributed through the valleys of the western depression of the Great Basin. Truckee Meadows, November 19.

173. **Spizella socialis arizonæ** Coues. WESTERN CHIPPING SPARROW.

San Diego. Rare migrant; perhaps a few breed in the vicinity; the first seen at San Diego, 1884, was on March 21, a pair.—L. B.

Poway. F. E. Blaisdell.—First, April 7, 1884.

Julian. N. S. Goss.—First, April 5, 1884.

San Bernardino. F. Stephens.—Tolerably common summer resident of the valley. Agua Caliente, March, 1884. Common. Seen April 1 and 8, 1886.

Henshaw, 1876. Pretty well diffused in the southern half of California; Santa Cruz Island, June.

Santa Catalina Island. F. Stephens.—August, 1886, common.

Santa Cruz. Joseph Skirm.—Tolerably common summer resident.

San Jose. A. L. Parkhurst.—March 28, 1884, first.

(Snow on the mountains at this time). First seen March 11, 1885, two specimens; common March 23; singing March 13; abundant summer resident.

Alameda and Contra Costa counties. W. E. Bryant.—Summer resident; tolerably common.

Haywards. W. O. Emerson.—Common summer resident. First seen April 6, 1885.

Berkeley. T. S. Palmer.—Common summer resident. First seen April 15, common April 21, 1885. In 1886 the first bird was seen April 4; common April 14.

Central California. L. B.—Moderately common summer resident in the valleys; very common in the mountains. Murphys, April 11, 1887, first, spring of 1877, early. Gridley, September 23, 1884, a large flock; October, 8, a small flock; October 20, a specimen. These are my earliest and latest records. Summit, September 25, 1885, still common; September 28, last.

Sebastopol. F. H. Holmes.—April 14, first; common April 17, 1885; abundant summer resident.

Marysville. W. F. Peacock.—April 10, first male; bulk arrived April 21, 1884. First seen March 17, common March 29, 1885. Tolerably common summer resident.

Chico. William Proud.—April 12, 1885, a pair.

Beaverton. A. W. Anthony.—Common summer resident. April 17, 1884, first; rare until the 25th, about which time the bulk arrived. First seen March 30, one bird; common May 15, 1885.

Suckley, 1860. Common throughout the two territories (Oregon and Washington).

British Columbia. John Fannin.—Common summer resident.

Henshaw, 1879. Common summer inhabitant of the eastern slope.

Hoffman. Generally distributed throughout the wooded districts of the upper half of Nevada.

Ridgway. In the interior in all the wooded districts, nowhere more abundant than in the Ruby Mountains in July and August.

Cooper, 1870. I found them wintering in the Colorado Valley in large numbers but not near San Diego.

174. Spizella breweri Cass. BREWER'S SPARROW.

Only five or six seen at San Diego, spring of 1884, when first collected by Col. N. S. Goss. March 8, 1884, apparently rare near the Pacific Coast in California. Found in summer only.—L. B.

Agua Caliente, Colorado Desert, San Diego County. F. Stephens.—March 25–28, 1884, abundant. Several seen April 10, 14 and 15, 1885.

Fort Tejon. Henshaw, 1876.—August, rather numerous.

Sacramento. Ridgway.—June 6, July 4, 1867, quite common in the bushy fields.

Newberry. Common in the Sacramento Valley.

Henshaw, 1879. East slope; a characteristic inhabitant of the sage brush.

Camp Harney. Bendire.—Common summer resident; breeds abundantly amongst the sage brush covered plains.

Hoffman. Quite common in the northern and middle areas, more particularly in the vicinity of settlements.

Ridgway. Throughout the entire extent of the Great Basin; everywhere one of the commonest birds of the open wastes. It arrived at Carson April 9, 1868.

Cooper, 1870. At Fort Mojave I found small flocks after March 20.

L. B.—Very common in Sierra Valley June, 1885, and a few were found July 1 on Castle Peak, Nevada County, up to 8,000 feet, found there in sage brush as is usual in the breeding season. But few were seen at Summit during the fall migration of 1885, though abundant August

10–13, 1882. Dr. Newberry said it was common in Sacramento Valley, but I seldom find it there, and only when migrating. Mr. Chas. A. Allen has collected it in Marin County, and Dr. Heermann at Tejon Valley and other parts of California, but probably mostly in the southeast part, as in Colorado Valley. The early explorers in their extensive marches seldom named the locality where their specimens were obtained, a distance of one hundred or even five hundred miles seeming, apparently, but a trifle to them, and there have been changes in the outlines of States and Territories since their time.

175. Spizella atrigularis (Cab.) BLACK-CHINNED SPARROW.

Poway. F. E. Blaisdell.—March 20, male specimen. May 4, nest with four small birds; nest in a small bush 1½ feet from the ground. May 19th, nest in a small solitary bush at the foot of a mountain; contents, five small birds. Same date, a nest two feet from the ground; eggs, four; the nest in a small bush. The eggs were light blue color same as those of *Chamæa fasciata;* one egg measured 17 mm. from end to end, and 40 mm. in circumference. The nests were composed of very small twigs and shreds of sage bark. Depth inside, one-half inch; inside diameter, two inches; outside, three inches. All three of the nests were of very nearly the same size and of the same material.

Santa Ana Plains, Los Angeles County, December 10–14, 1884, rather common.

San Bernardino. F. Stephens.—Tolerably common; breeds in the foothills.

Colton, April 28, 1884, a fine male shot by K. B. Herron; April 29, male shot by Charles W. Gunn.

[Mr. Stephens was the first to find it in the coast region of California, in 1883 or earlier.]

176. **Junco hyemalis** (Linn.) SLATE-COLORED JUNCO.

About thirty miles east of San Diego, January 24, 1884, a specimen taken and forwarded to the Smithsonian. It had another junco for a companion. A cold storm with snow occurred in these mountains the following day and this and other snowbirds were probably driven here by storms as I saw none in going to Campo January 22 and 23. On the 25th there was sleighing at Portland, Oregon, and a cold storm all along the coast. A female was collected by Mr. Emerson, March, 1880, at Haywards; and one by myself at Gridley.—L. B.

177. **Junco hyemalis oregonus** (Towns.) OREGON JUNCO.

San Diego, March 8, 1884, one bird; the only one seen near here during the winter. Campo. Winter, common.—L. B.

San Diego. F. E. Blaisdell.—November 22, 1885, a few in town. I also saw a pair at Poway, October 31, the first I have seen there in six years. The species breeds in the Volcan Mountains; in former winters, plentiful at Poway.

Volcan Mountains. W. O. Emerson.—Large flocks at all times.

San Bernardino Mountains. F. Stephens.—Common; breeds from 7,000 to 10,000 feet altitude.

Henshaw, 1876. Mountains near Fort Tejon, very abundant in August.

Santa Cruz. Joseph Skirm.—Common summer resident.

Cooper, 1870. Coast Mountains south of Santa Clara, many breeding in May. At San Diego I observed them until April 1. I have not determined its residence along the coast farther south than Monterey.

Alameda and Contra Costa counties. W. E. Bryant. Common winter visitant.

Sebastopol. F. H. Holmes.—Abundant winter visitant; the bulk left in the third week of March, 1885; April 9, last seen.

Olema. A. M. Ingersoll.—Common in April; May 1, last.

Beaverton. A. W. Anthony.—Abundant resident; first nest and eggs April 20.

Willamette Valley. O. B. Johnson.—Abundant during winter; a few remain to breed.

Seattle. O. B. Johnson.—April 14, 1884, nest, eggs nearly hatched; April 28, nest, eggs fresh; May 1, nest and five nearly fledged young.

Cooper, 1860. Common throughout Washington Territory, especially in winter; breeds about Puget Sound.

British Columbia. John Fannin.—Abundant resident.

Henshaw, 1879. Appears to pass the summer from about the latitude of Carson northward.

Camp Harney. Bendire.—Winter resident, retiring to the neighboring mountains in summer.

Ridgway. Very abundant in winter from the Pacific coast to the Wahsatch Mountains, but in summer has a more restricted distribution, being then confined to the coniferous forest region of the higher western ranges.

Summit. L. B.—November 13-16, a few; Big Trees, January 6, 1879, rather common; very common in summer in fir forests of the Sierra; nests on the ground, invariably, the nest varying much in composition, sometimes of grasses entirely, often of soap root *(Chlorogalum)* fibres almost entirely; rarely of shreds of bark of Libocedrus sparcely lined; very common in foothills and valleys of Central California in winter, as it unquestionably is in nearly all the State.

Chico. Wm. Proud.—October 1, 1884; first, two; abundant October 21.

Gridley, September 27, 1884, two; common October 1.—L. B.

Haywards. W. O. Emerson.—October 7, first (six); common at once.

Berkeley. T. S. Palmer.—Abundant winter resident. First seen September 27, 1885; common October 20; last seen April 20, 1886.

178. **Amphispiza bilineata** (Cass.) BLACK-THROATED SPARROW.

Probably never found in California west of the Sierra.—L. B.

Agua Caliente, San Diego County, Cal. F. Stephens.—Seen almost every day from March 26, to April 15, 1886.

Cooper, 1870. In winter they descended to the hills near the Colorado. They were never numerous.

Hoffman. Like *A. belli nevadensis*, this species occurs throughout the sagebrush regions in the northern and middle portions of Nevada, but in the southern regions was found in the elevated deserts, between Spring Mountain and the Colorado, containing an abundant growth of *Algarroba* and *Yucca baccata*.

Ridgway. Throughout the sagebrush country, the most desert tracts of which are its favorite abode. Abundant on the Carson desert, a summer sojourner, making its appearance in Truckee Valley, May 13, 1868. Juvenile specimens July 25, 1867, in Truckee Meadow.

179. **Amphispiza belli** (Cass.) BELL'S SPARROW.

San Diego. L. B.—Common resident.

Poway. F. E. Blaisdell.—Common resident.

San Bernardino. F. Stephens.—Rare summer resident of foothills.

Cooper, 1870. Numerous on San Nicolas Island, eighty miles from the mainland. They remain all the year at the same localities. At San Diego I found the young hatched May 18; think they are often earlier. Common in the chaparral of the Santa Clara Valley.

Henshaw. In the mountains near Fort Tejon it breeds abundantly up to 5,000 or 6,000 feet.

Contra Costa County. W. E. Bryant.—Rare. Breeds.

Nicasio. [North of San Francisco Bay]. C. A. Allen.

Foothills of Calaveras County, rare resident.—L. B.

Heermann. In the fall of 1851 I first noticed this species in the mountains bordering the Cosumnes River. We found it in great numbers between Kern River and the Tejon Pass and between the latter and Mojave River.

180. Amphispiza belli nevadensis (Ridgw.) SAGE SPARROW.

Desert side of San Bernardino Mountains. F. Stephens.

Henshaw, 1876. Kernville, a single specimen and saw others.

Ridgway. Most numerous in the valleys of the western depression, few being seen in the Salt Lake Valley where the *A. bilineata* was so abundant. In the neighborhood of Carson it was by far the most abundant bird of the open wastes and its abundance did not abate with the approach of winter. They began singing toward the last of February, and by the beginning of April the first eggs were laid. On the 9th several nests were found.

Henshaw, 1879. Breeds abundantly among the *Artemisia;* less abundant in northeastern California than southward.

Camp Harney. Bendire.—A not very abundant summer resident; none remain through the winter.

181. **Peucæa ruficeps** (Cass.) RUFOUS-CROWNED SPARROW.

Poway. F. E. Blaisdell.—Noticed January 27, 1884; April 16, I shot a female, which appeared to have a nest. I have seen no more than eight individuals here this season.

Cooper, 1870. I have only met with this species on Catalina Island in June, a few keeping about the low bushes.

Oakland and vicinity. W. E. Bryant.—Rare. Breeds.

Alameda and vicinity. A. M. Ingersoll.—Rare.

Nicasio. C. A. Allen.—Common every season in their favorite spots on the mountain sides.

Calaveras County. L. B.—December 13, 1878, altitude 3,000 feet; rare. I occasionally see it lower down in Calaveras County in winter.

Colfax. November 19, 1884, one specimen; several seen. This should carry it to the head of Sacramento Valley, only a few hundred feet above sea level, in suitable localities.

Cosumnes River. Heermann.—But one specimen. In the spring of 1852 in the mountains near the Calaveras River I found it quite abundant.

182. **Melospiza fasciata montana** Hensh. MOUNTAIN SONG SPARROW.

West Humboldt Mountains, Ruby Valley, Utah. Ridgway.

Camp Harney. Bendire.—The race found here, a resident, partly at least.

Henshaw, 1379. Several localities at the foot of the eastern slope, although it is along here that the change to *heermanni* is first indicated.

Hoffman. Rather common, more particularly so in the eastern and southern parts of Nevada.

Stockton. L. B.—September 29, 1881, first, a fine example; October 12, several. Gridley, September 30, 1884, appears to be a rather rare but regular winter visitant in the Sacramento Valley as far south as Stockton, coming in fall at about the same time that *M. fasciata guttata* does. Specimens collected by me at Marysville and Stockton in 1878 (and afterward, I believe), were identified by Prof. Ridgway as variety *fallax* Baird, which is now treated as variety *montana* Henshaw. I suppose Prof. Ridgway in giving the habitat of *montana* in his Manual of North American Birds, unintentionally omitted California, but I do not know such to be the fact. Perhaps the individuals found in California in winter cross the Sierra Nevada to winter in the milder climate of this State and in spring return to Nevada to breed. Very few song sparrows breed in the mountains of California.

183. Melospiza fasciata heermanni (Baird). HEERMANN'S SONG SPARROW.

An abundant resident in the tule marshes of the interior of California; also found frequently in thickets near water. Specimens I collected in January at Campo were not typical *heermanni*, taking the Stockton bird for such, but were nearer to it than to *samuelis*, taking Walter E. Bryant's diminutive Oakland *samuelis*-as a type for the latter. The local variation of the resident song sparrows of California is decidedly perplexing. Mr. Henshaw (1879), in an excellent article on the *Melospizæ*, says that it is upon a basis of size alone that *samuelis* can be separated from *heermanni*, *samuelis* being much the smaller. Charles H. Townsend found it breeding on McCloud River.

The type probably came from the southern part of the San Joaquin Valley, not far from Fort Tejon. The Bakersfield resident song sparrow is identical in size and color with that of Stockton and northward, or very nearly so.

San Bernardino. F. Stephens.—Common resident of the valley; rare resident of the foothills. Agua Caliente, San Diego County, east slope; common from March 18 to April 15, 1886.

West Humboldt Mountains, Truckee Reservation, Carson.—Ridgway.

184. Melospiza fasciata samuelis (Baird). SAMUELS'S SONG SPARROW.

San Quintin Bay, May, 1881, rare, not typical. San Rafael, moderately common May, 1885. San Diego, rare resident and quite like the above Lower Californian birds, differing from my types of *heermanni* and *samuelis*. Mr. Henshaw (1879) says it is *samuelis* alone that occurs along and near the Californian coast. I believe the type came from Petaluma.—L. B.

Volcan Mountains. Dr. Cooper.—Specimens obtained here by Mr. Emerson approach *heermanni*, but I think they will not do for it. They are identical with the Haywards birds.

Santa Cruz. Jos. Skirm.—Very common resident.

San Jose. A. L. Parkhurst.—Breeds.

Oakland. W. E. Bryant.—Common resident.

Nicasio. C. A. Allen.

Olema and Santa Cruz. A. M. Ingersoll.—Breeds.

Mouth of Eel River. C. H. Townsend.—A specimen December, 1885; the only one of the kind seen.

185. **Melospiza fasciata guttata** (Nutt.) RUSTY SONG SPARROW.

Henshaw, 1879. Variety *guttata* is characterized by a darker more rufescent type of color; the streaks on the dorsum are very indistinct, in some almost wanting. The typical home of this variety is the Columbia River region, coastwise, but long before reaching that point evidence is afforded by specimens of intermediate character of the changes to appear farther north. Variety *rufina* is simply *guttata* with the tendencies of the latter carried a step or two farther, with increase of latitude. The rufous of *guttata* in extreme cases becomes a reddish sepia brown; the size is somewhat larger, the bill rather more slender. This is *rufina* as found about Sitka and southward. Upon certain of the Alaskan Islands occurs *insignis*.

British Columbia. John Fannin.—Abundant resident.

Cooper, 1860. This species "*Melospiza rufina*" is a constant resident in the wild western portions of the Territory never ranging far from the thicket which contains its nest, or the house where it has found food and protection.

Suckley, 1860. Quite a common bird in the vicinity of Puget Sound, where it is a resident throughout the year.

Willamette Valley. O. B. Johnson.—Plentiful permanent resident.

Camp Harney. Bendire.—In December, 1875, I took two specimens, perhaps the true *guttata*.

West Humboldt Mountains. Ridgway.—Specimen, October 3.

Gridley. L. B.—September 23, 1885, first; September 30, common; these were what I consider intermediate specimens, resembling the song sparrow that breeds at Bear Valley, near Emigrant Gap, and at Sierra Valley. Gridley, October 14, much darker birds, tolerably common for the first time. All remain here in winter. Mr. Parkhurst sent me a specimen of typical *guttata* from San Jose collected in winter. I find it at Stockton, Murphys and Colfax in winter.

Nicasio. C. A. Allen.

Oakland. W. E. Bryant.—Tolerably common winter visitant.

186. Melospiza fasciata rufina (Bonap.) SOOTY SONG SPARROW.

Coast region of British Columbia, north to Sitka.

187. Melospiza lincolni (Aud.) LINCOLN'S SPARROW.

San Diego, tolerably common migrant. April 26, 1884, last seen—a fine male.—L. B.

Agua Caliente, San Diego County. F. Stephens.—Rather common winter resident.

Volcan Mountains. W. O. Emerson.—January 25, one only.

Oakland. W. E. Bryant.—Rare migrant.

Sebastopol. F. H. Holmes.—Rare winter visitant,? October 15, 1884.

Central California. L. B.—Rare summer resident of subalpine meadows; occasionally breeds in Calaveras County at 4,700 feet altitude. It was common in the winter of 1877–78 at Marysville, a colder winter than the average; presume some winter in suitable localities as far north as this every winter. Gridley, September 24, 1884, first; several. Tehachapi, April 4, 1889, first seen, many migrants.

British Columbia. John Fannin.—Common summer resident.

Bendire. Noticed in considerable numbers on their

way north in the spring of 1876 on Rattlesnake Creek. A few remain to breed.

Ridgway. During summer we found this species only in the elevated parks of the higher mountain ranges, but during its migrations it was very plentiful in the lower valleys. It arrived at Carson April 29, 1868.

L. B.—Butte Creek House, Cal., latitude 40° 10', altitude 5,600; nest and four nearly fresh eggs; nest of grass, stems and dead leaves lined with fine grass, placed on the ground on a cow chip in a very damp meadow near a creek. Two of the eggs were much paler than the others; all were greenish white, blotched, spotted and speckled with brown. Date of finding July 1st; female shot as she flew from the nest.

Mr. Brewster (Auk, 1889, page 90), refers one of the specimens I collected in the Victoria Mountains, Lower California, in February, 1883, to *M. lincolni striata*, Brewst., which he describes in the Auk of April, 1889.

188. **Melospiza lincolni striata** Brewst. FORBUSH'S SPARROW.

British Columbia.

189. **Passer domesticus** Linn. ENGLISH SPARROW.

Introduced about San Francisco Bay several years ago and now very abundant.* Reported breeding at Olema, spring of 1884 by Mr. Ingersoll; a few seen at Santa Cruz October 25, 1884, recent arrivals; the first seen by Mr. Skirm.

A few were brought from San Francisco to Stockton in the fall of 1883, and turned loose in the streets. They are multiplying rapidly and are already scattering

---

* *Phasianus torquatus* Gmel., *P. versicolor* Vieill. and *P. sœmmerringii* Temm., have been introduced into Western Oregon from China and Japan, and are thriving.

through the farming country about Stockton. I first noticed them at Sacramento, Marysville and Gridley in the spring of 1888. They nest early and often in California, beginning to breed early in March in the interior of the State.—L. B.

190. **Passerella iliaca** (Merr.) Fox Sparrow.

[A male taken near San Diego, January 3, 1888, by A. M. Ingersoll. See Proc. Cal. Acad., ser. 2, ii, 90.— W. E. B.]

191. **Passerella iliaca unalaschcensis** (Gmel.) Townsend's Sparrow.

British Columbia. John Fannin.—Common summer resident.

Washington Territory. Cooper, 1860. Only a winter resident. Most common in the interior, but in very cold weather seeks the coast.

Willamette Valley. O. B. Johnson.—Only a winter visitor.

Sebastopol. F. H. Holmes.—Tolerably common winter visitant; first seen September 24, 1884.

Gridley. L. B.—September 25, 1884, first; Big Trees, September 25, 1880; Summit, September 28, 1885, a dozen. These are my only dates of arrival from the north. I saw it up to March 15, 1884, at San Diego. It is a rare winter visitant to the low parts of California, but then more common in the foothills.

Berkeley. T. S. Palmer.—Common winter resident. Last seen February 14, 1886.

Alameda and Contra Costa counties. W. E. Bryant. Tolerably common winter visitant.

Henshaw, 1876. Early in October the mountains in the vicinity of Mt. Whitney began to be thronged with these birds.

San Bernardino. F. Stephens.—Rare winter visitant to valley and foothills.

Volcan Mountains. W. O. Emerson.—Very common along the creek in the thick brush until February 11. The snow then must have driven them down nearer the coast. Only one was seen after February 11.

Poway, F. E. Blaisdell.—A specimen November 17.

San Diego. L. B.—Rare winter visitant; probably goes but little south of San Diego.

192. Passerella iliaca megarhyncha (Baird). THICK-BILLED SPARROW.

San Diego. Rare winter visitor; April 1, 1884, last.— L. B.

Poway. F. E. Blaisdell. — Tolerably common in cañons; several specimens in April.

Julian. N. S. Goss.—April 17, I think it was nesting.

Volcan Mountains. W. O. Emerson.—Single birds seen February 9 and March 29. Both were in snow.

San Bernardino. F. Stephens.—Breeds at 8,000 feet altitude; rare.

Fort Tejon. Henshaw, 1876. Numerous enough in August.

Nicasio. C. A. Allen.

L. B.—Very common summer resident in fir forests of Calaveras and Alpine counties up to 7,000 feet or more; common in breeding season in Butte and Plumas counties, and probably farther north; tolerably common in the forest at Sierra Valley, Donner Lake and vicinity in summer; not at Alta and Colfax November 17–21, nor at Red Bluff in warm winter of 1884–85. I never see it in the lower foothills of Calaveras County in winter, though *P. unalaschcensis* is common there at that time. I found the first or earliest nest at Big Trees, June 14, 1879, June 12, 1880; nests here in snow bush or bear bush (*Ceanothus cordulatus*).

Ridgway. Met with only in the ravines of the Sierra Nevada near Carson and Washoe. It was entirely absent during the winter, not arriving from the south until about the 20th of April. It was particularly numerous in the lower portion of the mountains.

Henshaw, 1879. Present along much of the eastern slope and probably reaches quite to the Columbia River. I think it never descends into the lower valleys, which seem to be preferred by the allied form, *P. schistacea*.

193. Passerella iliaca schistacea (Baird). SLATE-COLORED SPARROW.

Poway. F. E. Blaisdell.—Specimen taken April 19 (identification correct).

Murphys. L. B.—January 4, 1879, one specimen (No. 77081, Smithsonian collection).

Ridgway. Carson.—First met during its northward migration, which began late in February or early in March; some few individuals having doubtless remained during the winter. The following September it was observed in the Upper Humboldt Valley.

Camp Harney. Bendire.—A common summer visitor; arrives about April 1. I have found some twenty nests within half a mile of the post.

194. Pipilo maculatus megalonyx (Baird). SPURRED TOWHEE.

San Diego, common resident.—L. B.

Santa Catalina Island. F. Stephens.—August, 1886, abundant.

Poway. F. E. Blaisdell.—Common resident.

Volcan Mountains. W. O. Emerson.—Seen in every walk.

San Bernardino. F. Stephens.—Common resident of valley, foothills and mountains. Agua Caliente, in cañons, probably resident.

Henshaw, 1876. Spread in great numbers over the southern half of California. On Santa Cruz Island one of the most numerous species.

Cooper, 1870. Santa Catalina and San Clemente Islands.

Alameda and Contra Costa counties. W. E. Bryant.—Common resident.

Berkeley. T. S. Palmer.—Common resident.

Central California. L. B.—Common resident below the pine forests; common summer resident in these to 7,000 or 8,000 feet altitude; partly migratory in upper Sacramento Valley. Specimens obtained here in winter inclining to *oregonus;* none typical of it.

Henshaw, 1879. The only form of the towhee met with by us; extremely numerous along the eastern slope as high as the Columbia River. I fully agree with Mr. Ridgway that specimens found along the eastern slope of the Sierra, are absolutely indistinguishable by external characters from *megalonyx*.

WallaWalla. Brewster (Bull. N. O. C., vol. vii, page 225). Nearly typical but showing slight approaches to var *oregonus*.

195. Pipilo maculatus oregonus (Bell). OREGON TOWHEE.

British Columbia. John Fannin.—Abundant resident.

Cooper, 1860. A constant resident of Washington Territory but does not frequent the coast except in winter.

Willamette Valley. O. B. Johnson.—A common constant resident.

Beaverton. A. W. Anthony.—Common resident. (Identification correct.)

Wilbur, Oregon. W. E. Bryant.—Adult and young specimens.

Nicasio. C. A. Allen.—(Perhaps in winter).

196. Pipilo chlorurus (Towns.) GREEN-TAILED TOWHEE.

Not seen by me at San Diego in winter nor in any part of California except one I shot at Marysville, February 12, 1878. This appeared to be a healthy bird and was shot while flying. I have not met with it in northern Lower California in May.—L. B.

Cooper, 1870. I found a few of them in winter in the Colorado Valley, and rather more at San Diego, but they left both places in March.

San Bernardino. F. Stephens.—Rare transient visitant to the foothills.

Henshaw, 1876. Breeding in the mountains near Fort Tejon; young taken August 1st. I did not find it lower than 5,000 feet.

Haywards. W. O. Emerson.—May 11, 1884, one—the first seen here. It was somewhat out of its bearings.

L. B.—Sierra of Central California, tolerably common summer resident as far north as 40° 10′, or more; usually nests here in "bear brush." The earliest record April 17, 1878. Last seen at Summit, September 25, 1885. I found it common, breeding in the sagebrush plains in Sierra Valley, as well as in the shrubbery in the forests.

Fort Klamath. Lieutenant Wittich.—Summer.

Henshaw, 1879. Rather numerous in summer along the eastern slope and reaches well into Oregon, if not indeed to the Columbia River.

Ridgway. On all the higher ranges from the Sierra Nevada to the Uintahs. Near Austin exceedingly abundant and breeding in the early part of July. It arrived at Carson April 25, 1868.

Camp Harney. Bendire—Moderately abundant during the summer.

Hoffman. First observed in the upper portion of the State, middle of May. In crossing the divide between Hot Spring cañon and Belmont, again seen.

197. Pipilo fuscus crissalis (Vig.) CALIFORNIAN TOWHEE.

San Diego. B. F. Goss.—March 16, full set of fresh eggs.

San Diego. L. B.—Common resident; shows slight approach to *albigula* of the Cape region; probably reaches to the Cape in modified form. Campo, winter; Bakersfield; Tehachapi, abundant.

Poway. W. O. Emerson.—Young out of nest April 14; Volcan Mountains, common.

Poway. F. E. Blaisdell.—Common resident.

San Bernardino. F. Stephens.—Common resident of valley and foothills.

Cooper, 1870. One of the most abundant and characteristic birds of California, residing constantly in all the lower country west of the Sierra Nevada and up to the summits of the Coast Mountains, 3,000 feet high.

Henshaw, 1876. Found by our parties in great abundance from San Francisco southward. At Santa Barbara in June young very numerous.

Alameda and Contra Costa counties. W. E. Bryant. Common resident.

Berkeley. T. S. Palmer.—Abundant resident.

Central California. L. B.—Abundant resident of valleys and foothills, especially the latter. Winters at Murphys, Colfax, Cherokee Flat and Red Bluff. In summer noticed on Stony and Salt Creeks, Colusa County, and on the east slope of the Coast Range in Mendocino County.

198. Pipilo aberti Baird. ABERT'S TOWHEE.

Cooper, 1870. Almost the exact counterpart, in the Colorado Valley, of *crissalis*. About April 1 I found many of their nests generally built in thorny shrubs.

Colorado Desert. F. Stephens.—A male secured [the most western record to my knowledge], March 22, 1886.

199. Cardinalis cardinalis (Linn.) CARDINAL.

I noticed six of these birds about the garden of Dr. O. Harvey at Galt, May 14, 1880. Mrs. Harvey informed me that a neighbor, Mrs. Long, had introduced them from Missouri. They seemed to be permanently located in the deciduous oaks and shrubbery in and near Dr. Harvey's garden. May 14, 1884, Miss Genevieve Harvey wrote: "We have not seen any cardinal birds for over a year; some persons think they were killed."

As I have since heard from them in the neighborhood of Galt and Newhope, I suppose they are thriving in the shelter of the thickets along the rivers.—L. B.

200. Habia melanocephala (Swains.) BLACK-HEADED GROSBEAK.

San Diego. L. B.—Rare summer resident; first male April 17; first female April 20, 1885.

Poway. F. E. Blaisdell.—April 6, first male; April 23, first female; commenced laying the last of April. First males, March 26, 1885.

Julian. N. S. Goss.—April 23, summer resident.

San Bernardino. F. Stephens.—Tolerably common summer resident of the valley; rare summer resident of the foothills.

Agua Caliente, San Diego County, Cal. F. Stephens. Every day from April 6 to 15, 1886.

Henshaw, 1876. We met with it at several places in southern California, where it is pretty well diffused.

Santa Cruz. Joseph Skirm.—Common summer resident.

San Jose. A. L. Parkhurst.—April 19, first, four or five in song.

Alameda and Contra Costa counties. W. E. Bryant. Summer resident, tolerably common.

Haywards. W. O. Emerson.—April 10, 1885 (♀), common April 21.

Berkeley. T. S. Palmer.—Common summer resident. First seen April 9, 1885; common April 30; last, August 26, 1885. In 1886, first seen April 23; common April 30.

Nicasio. C. A. Allen.—April 24, 1884, first.

Olema. A. M. Ingersoll.—April 22, 1884, first.

Stockton. L. B.—April 27, 1879, males common; April 27, 1880, first male; April 27, 1890, first male, singing; common in the valleys, foothills and fir forests of the Sierra; not found at Blood's, Hermit Valley or Summit, altitude 7,000 feet and upward, at least not so recorded, but common 1,000 feet lower; tolerably common about Donner Lake and Sierraville in summer.

Stockton. J. J. Snyder.—First seen April 19, 1885; common May 3.

Marysville. W. F. Peacock.—April 25, 1884, first male; bulk arrived May 11. First seen April 17, 1885; common May 1, 1885.

Murphys. J. P. Snyder.—First seen April 30, 1885; common May 3. Last seen Sept. 17, 1884.

Chico. Wm. Proud.—April 24, warm and sunny; one bird came in this morning. April 28, bulk arrived. First, April 11, 1885.

Sebastopol. F. H. Holmes.—First seen April 16, 1885; common April 30.

Willamette Valley. O. B. Johnson.—Common summer resident; breeding numerously.

Walla Walla. Dr. Williams.—First, April 27, 1885; young, June 6.

Cooper, 1870. Extending its migration as far as Puget Sound at least.

Henshaw, 1879. Carson, May 15, very numerous, and appeared to be mating.

Camp Harney. Bendire.—Rare summer visitor; begins to breed about June 1.

Hoffman. Found in Independence Valley and at Bull Run Mountain; was again noticed two miles south of Mineral Hill, but nowhere north of it.

Ridgway. Quite abundant in the fertile valleys and lower cañons along the entire route from Sacramento to the Wahsatch and Uintahs.

201. **Guiraca cærulea eurhyncha** Coues. WESTERN BLUE GROSBEAK.

Henshaw, 1876. Met with at several places in southern California where it is well diffused. Specimens at Los Angeles, Fort Tejon, Walker's Basin.

Stockton. John J. Snyder.—May 3, 1835, first; rare.

L. B.—Stockton, May 3, 1879, first males. May 6, 1878, first males; May 13, females. May 8, 1889, males and females. May 3, 1880, first males. Same date, many Gambel's geese going north and first western wood pewee. The blue grosbeak was common in summer about Stockton a few years ago. It did not appear to remain as late as September.

Hoffman. Met with in the valley north of Mount Magruder and on the western border of Deep Spring Valley along the base of the Inyo range, California.

Ridgway. Met with only at Sacramento where it was a very common bird in the bushy fields in the outskirts of the city.

Marysville. W. F. Peacock.—May 11, 1884, male and female; May 21, bulk arrived.

Chico. William Proud.—June 11, 1884, first male; rare this summer, usually common.

Newberry. ¡This pretty and musical little bird we found only on Pit River.

L. B.—Campo to San Diego, May 15 to 16, 1884, tolerably common, mated.

San Diego, April 26, 1884, first males, three; April 22, 1885; first male; April 23, several males.

Poway. F. E. Blaisdell.—Rare; April 24, 1884, first males—two; April 15, 1885, first.

Temecula. F. E. Blaisdell.—May 4, 1883, common.

San Bernardino. F. Stephens.—Rare summer resident of the valley.

Cooper, 1870. I noticed the first one at Fort Mojave May 6, and afterwards saw many more. I have also seen them at Los Angeles and Santa Barbara.

202. Passerina amœna (Say). LAZULI BUNTING.

San Diego. Rare summer resident; first male April 23; female, April 28. First seen April 16, 1885; males common April 20, no females.—L. B.

Poway. F. E. Blaisdell.—April 23, first; first eggs May 4, 1884. Last seen September 10. First, April 3, 1885; common April 15.

Poway. W. O. Emerson.—April 18, 1884, first.

Haywards. W. O. Emerson.—April 2, 1885, seven males; females arrived five days later. Common summer resident.

San Bernardino. F. Stephens.—Rare summer resident of the valley.

Henshaw, 1876. Found in great abundance in the sheltered valleys.

Santa Cruz. Cooper, 1870.—April 12, 1866.

Santa Cruz. Joseph Skirm.—Common summer resident.

San Jose. A. L. Parkhurst.—April 23, first—five or six males.

Stockton, Cal. J. J. Snyder.—April 19, 1885, first seen; common May 3; summer resident.

Alameda and Contra Costa counties. W. E. Bryant.—Tolerably common summer resident.

Nicasio. C. A. Allen.—May 7, 1884, first; April 27, 1876.

Olema. A. M. Ingersoll.—April 28, 1884, first.

Central California. L. B.—Tolerably common summer resident in the valleys and foothills; a few breeding at Big Trees and Blood's, the latter 7,200 feet altitude.

Marysville. W. F. Peacock.—Common summer resident. May 8, 1884, first; bulk arrived May 11. First seen April 12, 1885; common May 8.

Sebastopol. F. H. Holmes.—First seen April 21, 1885; common May 1. Abundant summer resident.

Chico. Wm. Proud.—May 7, 1884, first; bulk arrived May 16. First seen April 23, 1885; one male. Common summer resident.

Willamette Valley. O. B. Johnson.—Abundant during summer; breeding plentifully.

Walla Walla, W. T. Dr. Williams.—March 27, two males; common April 20, 1885. Common summer resident.

Cooper, 1860. Arrives at Puget Sound about May 15, and is abundant in open districts of the interior during summer.

Suckley, 1860. The specimen I shot at the Dalles was obtained from a flock of several hundred individuals which had just arrived from the south in spring.

Henshaw, 1879. A common summer resident about

Carson, but met with less frequently toward the north. Found to breed up to an altitude of 7,000 feet.

Camp Harney. Bendire.—A rare summer visitor; a pair or so breed.

Hoffman. Found in nearly all the fertile regions along the mountain slopes and valleys. Several specimens at Bull Run Mountain May 25; found later in the season in the elevated valleys near Mt. Nagle, Mt. Magruder, and Spring Mountain.

Ridgway. A very common species in all the fertile valleys, as well as in the lower cañons of the mountains.

Berkeley. T. S. Palmer.—An abundant summer resident. First seen April 27, 1885; common May 2. In 1886, first seen May 9; common May 15.

Beaverton, Oregon. A. W. Anthony. — First seen May 21, 1885; again May 23.

203. Calamospiza melanocorys Stejn. LARK BUNTING.

Near National City, six miles south of San Diego, May 6, 1884, thirty or forty males hovering and singing on the mesa.—L. B.

El Cajon. N. S. Goss.—May 16, 1884.

Poway. F. E. Blaisdell.—May 25, 1886.

Henshaw, 1875. Seen in Snake Valley, Nevada, by Dr. Yarrow, where it had not been noted before.

San Diego. L. B.—April 1, 1885, three male specimens from a flock of a dozen. April 16, 1885, flock in male and female plumage; April 30, 1885, mostly mated. Many persons about San Diego in the spring of 1884 noticed this, to them, strange bird, and as neither Drs. Heermann nor Cooper mentions its occurrence here, it may have recently extended its range to this part of California. I also saw a flock at Campo, and between Campo and San Diego, May, 1884.

204. **Piranga ludoviciana** (Wils.) LOUISIANA TANAGER.

San Diego. L. B.—Very rare migrant immediately on the coast. Very common between San Diego and Campo, May 15–16, 1884, the sexes being about equal. Nearly all passed by San Diego in their journey north prior to May 9, 1885. None spend the winter in California, being then south of this State.

Poway. F. E. Blaisdell.—April 29, 1884, first, two seen; May 22, a flock. Common in April and May, 1883. Common May 3, 1885.

Julian. N. S. Goss.—April 22.

San Bernardino. F. Stephens.—Rare summer resident of the mountains; tolerably common migrant in valley and foothills.

Henshaw, 1876. Near Santa Barbara in July, several, feeding their young. Here and elsewhere in southern California, rare.

Sebastopol. F. H. Holmes.—May 1, 1885.

Alameda and Contra Costa counties. W. E. Bryant. Rare migrant.

Chico. William Proud.—May 3, 1885.

Berkeley. T. S. Palmer.—Common in spring of 1883; rare in 1885; last seen September 11, 1885.

Ukiah. George E. Aull.—Rare summer resident.

Stockton. L. B.—May 7, 1880, first male; first males in 1885, May 3; May 5, 1889. Galt, Sacramento County, May 13, 1880, common—mostly males. Gridley, July 8, 1889, a male, feeding on mulberries. Big Trees, May 25, 1880, both sexes common and mated. It is common in the firs of the Sierra from latitude 38° northward, where it generally nests well up in conifers, though I have known it to nest twelve or fifteen feet from the ground in dogwood *(Cornus Nuttallii)*.

Gridley, July 28, 1885. I was surprised to find these in the valley in July, where it is known only as a migrant. Mr. John J. Snyder shot a fine, fat male adult at Stockton July 19, 1885.

Murphys. John J. Snyder.—October 4, 1885, still to be seen.

Beaverton, Oregon. A. W. Anthony.—May 14, 1885, six specimens; common May 20; not uncommon summer resident.

Willamette. O. B. Johnson.—Common summer resident.

Cooper, 1860. Arrives at Puget Sound about May 15; common summer resident.

British Columbia. John Fannin.—Common summer resident. (Burrard Inlet, May 12, 1885.)

Henshaw, 1879, Summer visitor of the mountains as high up at least as the Columbia River. A nest on the extremity of a limb of cottonwood about 15 feet from the ground.

Hoffman. Rather common in the timbered areas along the water courses in June, though later it was found in the wooded regions of the interior.

Ridgway. In May soon after their arrival from the south these tanagers were very numerous in the rich valley of the Truckee near Pyramid Lake. Very few were seen later in the season, nearly all having departed for the mountain woods. During July and August, common, east slope of Ruby Mountains.

205. Piranga rubriceps Gray. GRAY'S TANAGER.

Colombia, South America. Accidental in California. (Dos Pueblos, Santa Barbara county. *Cf.* Bryant, Auk, 1887, p. 78.)

206. Piranga rubra cooperi Ridgw. COOPER'S TANAGER.

Cooper, 1870. This beautiful bird I found quite common at Fort Mojave after April 25, 1861.

207. Progne subis hesperia Brewst. WESTERN MARTIN.

A dozen or more of both sexes were temporarily sojourning at San Diego April 28, during a cool rainstorm. It does not appear to breed on the coast about San Diego.—L. B.

Poway. F. E. Blaisdell.—Twelve miles from the coast. First seen May 1, 1884, a few only; common in the spring of 1883.

Little Santa Maria Valley. W. O. Emerson.—April 4, 1884, one only.

Julian. N. S. Goss.—April 4, 1884.

San Bernardino. F. Stephens.—Rare summer resident in the mountains; rare migrant in the valley.

Santa Cruz. Joseph Skirm.—Common.

San Jose. A. L. Parkhurst.—First seen May 3, 1884, two or three; they did not remain.

Contra Costa County. W. E. Bryant.—Rare summer resident.

Olema. A. M. Ingersoll.—First seen May 8, 1884; breeds.

Stockton. L. B.—March 1, 1879, one bird; again seen March 3; both sexes common March 12; peaches partly in flower March 12; first male, March 5, 1886. Common summer resident here and in many localities in Central California below fir forest, where it is very rare.

Marysville. W. F. Peacock.—March 17, first male; bulk arrived by March 21, 1884.

Marysville. Frank Manning.—March 17, 1884, a pair; no more seen for three days.

Chico. Wm. Proud.—First seen May 22, 1884; four birds at evening.

Cooper, 1860. I never saw this bird in Washington Territory.

Suckley, 1860. I obtained at Fort Steilacoom a specimen of *Progne*.

Henshaw, 1879. Colonies encountered at numerous localities among the pine woods of the mountains, where they are quite local.

Ridgway. Rare east of the Sierra Nevada. In Carson it was not common, while in Virginia City but a single individual was seen June 18, 1868.

San Jose. A. L. Parkhurst.—Arrived April 9, 1885.

Stockton. John J. Snyder.—March 6, 1885, four first seen; next seen March 7 and 8; common March 21.

Murphys. Jacob P. Snyder.—April 25, 1885, two, first seen; next seen April 29. These two lots are all I have seen this year. They were quite numerous in corresponding periods of previous years. [Murphys, March 15, 1878.]

Sebastopol. F. H. Holmes.—First seen April 17, 1885; next seen April 30; rare; breeds.

Marysville. W. F. Peacock.—March 22, 1885, first seen; next seen March 28; common March 29. Common; breeds.

Chico. Wm. Proud.—April 13, 1885, first seen; heading direct for the old breeding place.

Sierra Valley. L. B.—June 18-21, common; breeding several little martin houses recently erected for their use; not known to do so elsewhere on the Pacific Coast.

208. Petrochelidon lunifrons (Say). CLIFF SWALLOW.

One of the most abundant species in California. First seen at San Diego April 5, 1884, arriving in force; very numerous. First seen March 25, 1885; common April 10. They breed under eaves of buildings here as they now usually do in towns of California, though many still nest in cliffs in different parts of the State. Stockton, March 8, one; March 10, 1879, already common; apricots and peaches in flower; season about as early as usual. Murphys, March 15, 1878; common.—L. B.

Poway. F. E. Blaisdell.—Common summer resident. First seen April 5, 1884; last seen, Sept. April 22, 1885, first.

Volcan Mountains. W. O. Emerson.—April 2, 1884, about a dozen. Santa Maria. April 4, building in an old adobe house. Poway. April 12, nesting.

Julian. N. S. Goss.—April 10.

San Bernardino. F. Stephens.—Abundant summer resident of valley and foothills up to April 2, 1884. When I left for Tucson I saw no cliff swallows, barn swallows or purple martins. They are later migrants than the white-bellied and violet-green swallows which had been going and coming for weeks.

Agua Caliente, San Diego, Cal. F. Stephens.—Seen every day from April 8 to 14, 1886.

Santa Cruz. Joseph Skirm.—Common summer resident. First seen March 28, 1881; March 17, 1882; April 1, 1883.

San Jose. A. L. Parkhurst.—April 5, 1884, arrived in flocks. March 22, 1885, first; common April 1, (W. O. Emerson; hundreds on telegraph wires Sept. 9, 1886.)

Alameda and Contra Costa counties. W. E. Bryant.—Abundant summer resident.

Haywards. W. O. Emerson.—April 6, 1885; first, and common.

Stockton. J. J. Snyder.—March 29, 1885, first; common April 26.

Berkeley. T. S. Palmer.—Common summer resident. Last seen August 5, 1885. In 1886, arrived April 3.

Nicasio. C. A. Allen.—First seen April 20, 1884.

Sebastopol. F. H. Holmes.—First seen March 31, 1885; common April 7.

Marysville. W. F. Peacock.—March 23, 1884, first; bulk arrived April 8. March 18, 1885; common March 28.

Murphys. J. J. Snyder.—Two seen March 22; again March 26. (March 16, 1885; common March 29.—J. P. S.)

Beaverton. A. W. Anthony.—I think it came March 26 with *T. thalassina*, but not recognized until common.

Willamette Valley. O. B. Johnson.—Abundant in summer; breeding chiefly under eaves.

Cooper, 1860. At Olympia a few flying about the streets in July, rather scarce north of the Columbia River.

Suckley, 1860. Fort Dalles. Moderately abundant; makes its appearance in spring simultaneously with *Tachycineta bicolor* and *T. thalassina*, but is not so numerous.

British Columbia. John Fannin.—Summer resident east of Cascades.

Camp Harney. Bendire.—One of the most abundant summer residents.

Hoffman. Usually abundant in the vicinity of rivers, streams and even large springs.

Ridgway. Noticed along every portion of our route across the Great Basin, especially in the vicinity of rivers or lakes or at settlements whether large or small,

Cooper, 1870. In June I saw a flock of these birds busily catching young grasshoppers on the dry hillside where these insects were swarming.

Salt Spring Valley. (Calaveras County.) L. B.— September 13, 1884, a few about the reservoir.

209. Chelidon erythrogaster (Bodd.) BARN SWALLOW.

National City. G. Holterhoff.—One seen March 26, 1884.

Soledad. L. B.—Rare summer resident. Arrived April 4, 1885.

San Diego. One seen March 26, 1884. A few only seen afterward, especially after cool storms with snow in the neighboring mountains. A few April 8, 1885.

Poway. F. E. Blaisdell.—Usually a common summer resident; none this spring. (1884.)

San Bernardino. F. Stephens.—Rare migrant in the valley.

Henshaw, 1876. Far less numerous than the cliff swallow, though on the coast at least it is not rare. Santa Cruz Island, breeding.

Santa Cruz. Joseph Skirm.—Common summer resident. March 26, 1881, first; March 14, 1882, first; March 30, 1883, first.

San Jose. A. L. Parkhurst.—Common summer resident. April 7, 1884, first; two seen. In 1885, first seen March 22; common April 7.

Alameda and Contra Costa counties. W. E. Bryant. —Tolerably common summer resident.

Haywards. W. O. Emerson.—Common summer resident. Arrived March 28, 1885, and was common at once.

Berkeley. T. S. Palmer.—Tolerably common summer resident. Arrived in 1886, on March 25; next seen April 3; common May 11.

Nicasio. C. A. Allen.—March 31, 1884, first.

Olema. A. M. Ingersoll.—Common summer resident.

Murphys. J. J. and J. P. Snyder—September 10, 1884, still common, arrived March 16; common March 29, 1885. Several seen March 24, 1886; again on the 26th. [Murphys, March 15, 1878.]

Stockton. Dr. Hudson.—March 19, 1870, arrived and began to repair old nests.

Stockton. J. J. Snyder.—Rare in 1885. Arrived March 17.

Stockton. L. B.—March 9, 1879, first; one only. The weather became cool on this day and no more were seen until the 20th. I saw two about a barn near Stockton, March 13, 1885. It is at least common in Central California up to about 3,000 feet, a few breeding about dwellings above that height. It leaves Central California from October 1st to 15th.

Soquel (near Santa Cruz). W. O. Emerson.—September 14, 1884, nest with young half fledged.

Marysville. W. F. Peacock.—Common summer resident. March 27, 1884, first; April 8, bulk arrived. In 1885, six birds arrived April 5; common April 11.

Sebastopol. F. H. Holmes.—Common summer resident. Arrived March 22; common April 12, 1885.

Beaverton, Or. A. W. Anthony.—May 2, 1884, first, and occasionally afterward. The only one seen in 1885 was on May 12.

Walla Walla.—Dr. Williams.—Breeds; arrived May 20; ten specimens; common May 30, 1885.

Cooper. Seems to be limited by the Columbia River. as I have seen none at Puget Sound or more northern places.

British Columbia. John Fannin.—Abundant summer resident.

Burrard Inlet. First seen March 10, one male; common April 2, 1885.

Camp Harney. Bendire.—A few pairs breed about the buildings of the Post. They arrive about the same time (May 1), but remain much longer than the cliff swallow which leaves about the middle of August.

Hoffman. I noticed these birds north of Battle Mountain during the last days of May.

Ridgway. Although inhabiting the same localities as the cliff swallow, everywhere much less numerous. Several nests were found in caverns on the eastern side of Ruby Mountains. It arrived at Carson, April 8, 1868.

Cape Flattery Light. Alexander Simpson, keeper.—Arrived May 7, 1885.

Whidley Island, W. T. Lawrence Wessel.—Arrived April 24, 1885.

210. **Tachycineta bicolor** (Vieill.) TREE SWALLOW.

San Diego. Of occasional occurrence in winter; not at Campo and mountains sixty miles south in May, 1884. Perhaps none breed as far south as San Diego where it was abundant in February, but not seen after the first of April.—L. B.

San Bernardino. F. Stephens.—Tolerably common migrant in the valley. Agua Caliente, several seen, the latter migrating. (?)

Santa Cruz. Joseph Skirm.—Common summer resident.

San Jose. A. L. Parkhurst.—Abundant summer resident. February 24, 1884, four birds, first. Weather very warm for date. Arrived in 1885; February 20, two birds; common March 13. W. O. Emerson. September 9, 1886, hundreds of cliff and white bellied swallows together on telegraph wires.)

Sebastopol. F. H. Holmes.—Arrived March 9; common March 15, 1885.

Marysville. W. F. Peacock.—Abundant summer resident. February 20, 1884, both sexes; February 25 the bulk arrived. Arrived in 1855, January 16; common, January 28; January 21, 1886.

L. B.—Marysville, February 1; 1878, first; common soon afterward; weather not unusual.

Chico. William Proud.—Six birds seen February 16, 1885.

Haywards. W. O. Emerson.—Rare summer resident; April 8, 1885.

Stockton. L. B.—December 5 and 6, 1879, about a

hundred; also December 5, 1878, a large flock. January 18, 1855, I saw hundreds or thousands in a drive of six miles in the country. The next day being densely foggy, none were seen. It is a very common summer resident of Central California, more especially in the valleys and foothills, nesting about buildings as well as in knot-holes of trees in the country. I saw a pair carrying grass into a knot-hole at Stockton as early as March 25, 1879.

Stockton. J. J. Snyder.—Arrived January 18, two birds; common January 30, 1885.

Beaverton, Or. A. W. Anthony.—Common summer resident. Arrived April 4, two birds; common April 15, 1885.

Willamette Valley. O. B. Johnson. — Abundant; nesting in holes in trees.

Cooper, 1860. Common in the western portions of the Territory.

Suckley, 1860. Obtained both at Fort Steilacoom and Fort Dalles.

British Columbia. John Fannin.—Summer resident; more abundant east of the Cascades. (Burrard Inlet, arrived March 13; common March 29, 1885.)

Henshaw, 1879. More or less abundant inhabitant of the eastern slope; not met with farther north than northern California. In the mountains near Camp Bidwell it was numerous enough the last of July, at which time the pairs all had young as was shown by frequent visits to woodpeckers' holes in the aspens.

Ridgway. Among the cottonwoods of the lower Truckee near Pyramid Lake in May, more abundant than elsewhere, and every knot-hole or other cavity among the trees seemed to have been taken possession of by a pair. At Carson they were quite numerous and built their nests under the eaves, behind the weather boarding, or about the porches of dwellings or other buildings. Three specimens, Carson, March 30, 1868.

Coupeville, Island County, W. T. Lawrence Wessel. April 24, 1885.

Nicasio, Marin County, Cal. C. A. Allen.—April 2, 1876.

211. **Tachycineta thalassina** (Swains.) VIOLET-GREEN SWALLOW.

Not noticed immediately on the coast about San Diego where trees are scarce.—L. B.

Poway. F. E. Blaisdell.—Not seen this spring; usually common in spring and summer.

Julian. N. S. Goss.—April 14, 1884.

Volcan Mountains. W. O. Emerson.—St. Patrick's day, early in the morning, five clad in richest green, flying about for three hours; snow two inches deep. Also a flock March 20. April 1 they were about oak trees looking for holes for summer homes.

San Bernardino Mountains. F. Stephens.—Common summer resident, nesting in knot-holes, etc., in trees. It is a common migrant through the valley.

Agua Caliente, Colorado Desert. F. Stephens.— Abundant; perhaps migrants. Abundant March 18, 1886, and seen most every day up to the time of leaving the locality, April 15.

Henshaw, 1876. Along the coast, very numerous; abundant in September in the high meadows near the base of Mount Whitney.

Santa Cruz. Joseph Skirm.—Common summer resident, arriving March 28, 1881, March 21, 1882.

Contra Costa County. W. E. Bryant.—Rare summer resident.

Sebastopol. F. H. Holmes.—One specimen taken January 1, 1885; first arrival February 22; common March 1.

Nicasio. Chas. A. Allen.—March 15, 1884, first.

Beaverton, Or. A. W. Anthony.—Common summer resident. March 26, first; bulk arrived April 5, 1884. In 1885, first arrived March 13, five birds; common March 25.

Wilamette. O. B. Johnson.—Abundant; nesting in knot-holes and crevices about buildings; decidedly the most familiar of the three species of swallows.

Coupeville, Island County, W. T. Lawrence Wessel. April 2, 1885.

Suckley, 1860. Abundant throughout the interior of Oregon and Washington Territory. I have observed it arrive at Puget Sound about the 10th of May.

British Columbia. John Fannin.—Common summer resident. Burrard Inlet. First seen March 13; common March 29, 1885.

Henshaw, 1879. Extremely abundant summer visitant in certain portions of eastern California and western Nevada, as for instance at Pryamid Lake. Thousands resort to the niches and holes in the faces of the rocks for nesting sites. In the mountains where it is also abundant it selects for this purpose the deserted holes of woodpeckers, giving preference to those in oaks.

Bendire. Noticed on Bear Creek, Blue Mountains, summer of 1876.

Ridgway. The beautiful violet-green swallow was first seen at Pyramid Lake in May. They were very abundant and frequented chiefly the cliffs of calcareous tufa, where they were observed to enter the fissures of the rock to their nests within. In July we saw it again among the limestone walls of the eastern cañons of the Ruby Mountains where it also nested in the crevices on the face of the cliffs.

L. B.—My latest Californian record is Big Trees, September 25, though Dr. Cooper saw a large flock at

Santa Cruz October 5. I have never found it breeding in the valleys of the northern half of the State where it is rare even as a migrant. Murphys, breeding in cliffs with *Micropus melanoleucus.*

212. Stelgidopteryx serripennis (Aud.) ROUGH-WINGED SWALLOW.

Poway. F. E. Blaisdell.—Usually common in spring; not seen spring of 1884.

San Jose. A. L. Parkhurst.—Common summer resident. First seen March 8; common March 20, 1885.

San Bernardino. F. Stephens.—Rare migrant through the valley.

Agua Caliente. F. Stephens.—Several seen March 18, 1886.

Southern California. Henshaw, 1876.—Occurs commonly, its distribution being regulated only by the presence or absence of suitable localities.

Santa Cruz. Joseph Skirm.—Common summer resident.

Santa Cruz. A. M. Ingersoll.—Eggs taken here.

Contra Costa County. W. E. Bryant.—Summer resident.

L. B.—I have seen a few individuals at each of two localities in Calaveras County, and have not recognized it elsewhere in California.

Ridgway. Sacramento, June 6 to July 4, 1867, common.

Newberry. Found in California, and as far north as the Columbia River.

Cooper, 1860. Common about the sandy cliffs and inlets of this coast. It arrives near the Columbia River in May and remains until the middle of August.

Suckley, 1860. Rather abundant both in Oregon and in Washington Territory.

British Columbia. John Fannin.—Common summer resident. March 20, 1885, only a few seen this year.

Henshaw, 1879. Present along much of the eastern slope.

Hoffman. I noticed these birds along the banks of the Humboldt River, north of Battle Mountain, during the last days of May.

Ridgway. Next to the cliff and white-bellied swallows this was the most abundant species of the family. Arrived at Carson April 15, 1868.

Cooper, 1870. I saw them at Fort Mojave on the 22d of February, but I have seen them at San Diego November 9 and January 27, so that if they do not winter in the State they do not go far beyond it.

Whidby Island, W. T. Lawrence Wessel.—April 2, 1885.

Walla Walla, W. T. Dr. Williams.—May 25, three birds; again May 26; still present August 9, 1885.

213. **Clivicola riparia** (Linn.) BANK SWALLOW.

San Bernardino. F. Stephens.—Rare migrant through the valley.

Santa Cruz. Joseph Skirm.—Common summer resident.

Santa Cruz. A. M. Ingersoll.—I have collected the eggs here.

Newberry. Not uncommon throughout California. We occasionally saw this and the next species occupying their characteristic burrows. We sometimes probably confounded them as they are only to be distinguished by careful inspection.

Heermann. Abundant. (At Sacramento City?)

Contra Costa and Alameda counties (precise localities not named). W. E. Bryant.—Rare summer resident.

Nicasio. C. A. Allen.—March 19, 1876.

Sebastopol. F. H. Holmes.

British Columbia. John Fannin.—Summer resident east of Cascades; rare west of Cascades.

Henshaw, 1879. Present in Nevada and eastern California, but apparently less numerous than the rough-winged swallow.

Bendire. Breeds in large numbers on one of the islands in Malheur Lake.

Ridgway. Distribution the same as that of the rough-winged species but everywhere less abundant. Truckee Reservation in May, a few individuals; Sacramento City, common.

214. Ampelis garrulus Linn. BOHEMIAN WAXWING.

British Columbia. John Fannin.—East of Cascades and northward. Accidental west, though sometimes appearing here in flocks in winter.

Fort Walla Walla. Captain Bendire (Bull. N. O. C., 7).

Camp Harney. Captain Bendire (Birds S. E. Or.) Just noticed November 23, 1875. Quite a number were secured between November and March 1.

Fort Mojave, latitude 35°. Cooper, 1870. January 10. The specimen obtained was a straggler.

Willamette Valley. O. B. Johnson.—I obtained a pair of these beautiful birds during a snow-storm in January, 1876, at Forest Grove.

215. Ampelis cedrorum (Vieill.) CEDAR WAXWING.

San Diego. Winter of 1883–84, common until April 25; feeding on berries of the pepper tree *(Schinus molle)* L. B.

San Diego. Carl H. Danielson.—Several seen May 14, 1884. None seen this winter (1885).

Poway. F. E. Blaisdell.—May 18, a large flock, and previously in winter.

San Bernardino. F. Stephens.—A very irregular winter visitant to the valley; at times common.

Los Angeles. N. S. Goss.—March 6.

Santa Cruz. Joseph Skirm.—Large flocks sometimes come here, but stay only a short time.

San Jose. A. L. Parkhurst.—Several seen February 15; very large flocks February 21; a large flock March 1.

Gridley. L. B.—December 11 a flock feeding on wild grapes. I have seen flocks at Murphys in our coldest weather.

Chico. William Proud.—May 15, 1884, a small flock.

Beaverton. A. W. Anthony.—Common summer resident. First seen May 22; common by June 7, 1884.

Willamette Valley. O. B. Johnson.—Abundant summer resident.

British Columbia. John Fannin.—Common summer resident.

Ridgway. Upper Humboldt Valley.—Common in September.

San Jose. A. L. Parkhurst.—March 30, 1885, a small flock; May 10, large flocks; the last seen.

Oakland. W. E. Bryant.—A flock of nine in March, 1885.

Berkeley. T. S. Palmer.—Accidental visitant.

Beaverton. A. W. Anthony.—April 20, 1885, first seen; common April 24. Common summer resident; breeds.

Burrard Inlet. John Fannin.—First seen May 24, 1885; next seen May 26; common June 2. Breeds.

216. **Phainopepla nitens** (Swains.) PHAINOPEPLA.

San Diego. L. B.—April 24, 1884, male and female seen. I find the species in the foothills of San Joaquin and Calaveras counties quite as common in winter as in summer, though its presence depends much on the abundance or scarcity of mistletoe berries during the winter months more than on the temperature.

Poway. F. E. Blaisdell.—Common summer resident; first male seen May 6, 1884, the first female May 11. The latter and her mate had commenced a nest in a small oak tree May 11. It does not remain long after breeding; last seen August 2.

San Bernardino. F. Stephens.—Tolerably common summer resident of the valley and foothills.

Colorado Desert. F. Stephens.—March 19-26, 1886.

Marysville. W. F. Peacock.—March 8, 1884, one specimen. It does not breed here.

Chico. William Proud.—Two specimens; [information verbal; date forgotten].

Fort Crook. Baird, Brewer and Ridgway.—April, 1860, found by Feilner but has not been met with near the coast so far north.

Hoffman. It was observed in the valley at the northern slope of Mt. Magruder, on the eastern timbered foothills of the Myo Range, west of Columbus, and again at Spring Mountain, near the old " Spanish Trail," though only at rare intervals.

Cooper, 1870. On the Colorado I found them numerous especially in winter, and they do not migrate much south of latitude 35°. I also found them rather common along the Mojave River in December.

217. Lanius borealis Vieill. NORTHERN SHRIKE.

British Columbia. John Fannin.—Resident, not common.

Dr. Cooper, 1860. The northern shrike is only a winter resident in Washington Territory, appearing along the coast in November and remaining until March.

Willamette Valley. O. B. Johnson.—Quite common resident.

Sheep Rock, Northern California. Lieutenant John Feilner.—May 16, 1860, "might be said to be numerous."

Nicasio. C. A. Allen.—Winter.

Marysville. L. B.—Winter of 1878, six specimens.

Carson. R. Ridgway.—(Bull. Essex Inst. 1874.) Not common.

Camp Harney. Bendire.—Rare winter visitor.

218. Lanius ludovicianus gambeli Ridgw. CALIFORNIA SHRIKE.

San Diego. L. B.—Common resident.

Poway. F. E. Blaisdell.—Resident; first nest April 6. I usually obtain its eggs about March 12.

San Bernardino. F. Stephens.—Common resident of the valley.

Agua Caliente. Not common. Eight specimens seen from March 24, to April 14, 1886. Resident?

Santa Cruz. Joseph Skirm.—Common.

San Jose. A. L. Parkhurst.—February 27, 1885, full grown brood of young.

Alameda and Contra Costa counties. W. E. Bryant.— Tolerably common resident.

Berkeley. T. S. Palmer.—Tolerably common resident.

Sebastopol. F. H. Holmes.

Central California. L. B.—Common resident; generally distributed below 2,500 feet; wintering at Murphys, Colfax, Red Bluff, and probably farther north. Rarely seen above 2,500 feet on the west slope, in other words, rarely found in the pine or fir forests, breeding however in Sierra Valley and other parts of the east slope up to 5,000 feet or more, as do several species, *Carpodacus mexicanus frontalis*, *Icterus bullocki*, and others, which find summer homes much higher on the east than on the west slope. Although common, never numerous.

Henshaw. 1879. East slope, most numerous in summer, chiefly in the lower foothills.

Camp Harney. Bendire.—A common summer visitor and generally distributed. They arrive here about the 20th of March.

Hoffman. Over nearly the whole of the area from Bull Run Mountain southward as far as Belmont, and westward toward the Inyo Mountain range, California.

Ridgway. Scarcely a locality was visited where this shrike was not found in greater or less plenty, both in the Sacramento Valley and eastward of the Sierra Nevada.

Cooper, 1870. Many about Fort Mojave in winter.

219. Vireo olivaceus (Linn.) RED-EYED VIREO.

Walla Walla. J. W. Williams.—June 4 and June 24, 1885, I took specimens; in all six specimens. It was not very rare, and I could have taken as many more had I been so disposed.

[Dr. Williams correctly identified these and several other birds which he kindly sent to me.]

British Columbia. John Fannin.—Summer resident; not common.

220. Vireo flavoviridis (Cass.) YELLOW-GREEN VIREO.

A female taken at Riverside, Cal., by W. W. Price, Auk, V, 210.

221. Vireo gilvus (Vieill.) WARBLING VIREO.

San Diego. Rare summer resident.—L. B.

Poway. F. E. Blaisdell.—Common summer resident; April 2, already common; nest and eggs June 11, 1883. First arrival, March 29, 1885; common, April 22.

San Jose. A. L. Parkhurst.—First arrival March 15, 1885. Common summer resident.

Live Oak Springs, San Diego County. W. O. Emerson.—April 5.

San Bernardino. F. Stephens.—Common migrant in foothills and mountains. Agua Caliente, one was shot, probably migrating. A few seen March 19 to April 12, 1886.

Henshaw, 1876. Occurs commonly in California, inhabiting the deciduous trees of the low districts, and extending upward at least 10,000 feet.

Santa Cruz. Joseph Skirm.—Common summer resident.

Haywards. W. O. Emerson.—First arrival March 28, 1885; common April 18.

Oakland and vicinity. W. E. Bryant.—Common summer resident; first seen April 3; usually arrives about March 22. (March 12, 1885. ♂.)

Berkeley. T. S. Palmer.—Abundant summer resident. April 4, 1885; common April 15. In 1886 it arrived March 28; common April 3.

Marysville. W. F. Peacock.—First arrival April 5, 1885; common April 12. Common summer resident.

Sebastopol. F. H. Holmes.—First arrival April 21, 1885. Common summer resident.

Nicasio. C. A. Allen.—First seen April 24—March 28, 1876.

Stockton. L. B.—Several specimens of both sexes March 25, 1879. Common summer resident in nearly all altitudes.

Beaverton. A. W. Anthony.—Common summer resident. First seen April 27; heard singing a few days later; bulk arrived by May 12.

Puget Sound. Cooper.—Common after about May 15.

British Columbia. John Fannin.—Summer resident east of the Cascades.

Camp Harney. Bendire.—Seen on but two occasions, in June, 1876.

Henshaw, 1879. Numerous in the mountains near Camp Bidwell.

Ridgway. The habits and notes of the western birds of this species are in all respects like those of the eastern ones.

222. Vireo solitarius (Wils.) BLUE-HEADED VIREO.

Ridgway. Met with only during its autumnal migrations, when it seemed to be not uncommon in the month of September among the cañon thickets of the western slope of the Clover Mountains. A single specimen was also shot on the eastern slope of the West Humboldt Mountains in September of the preceding year (1867).

British Columbia. John Fannin.—Summer resident; not common.

223. Vireo solitarius cassinii (Xantus). CASSIN'S VIREO.

Tia Juana. N. S. Goss.—Fifteen miles south of San Diego, March 20.

San Diego. L. B.—April 20, 1884, first seen, one male; in breeding season not seen at San Diego.

Poway. F. E. Blaisdell.—A specimen April 16, 1884.

San Bernardino. F. Stephens.—Very rare migrant in the valley. Tolerably common summer resident of the foothills and mountains.

Agua Caliente. One seen April 7; three, April 12, and one April 13, 1886.

Henshaw, 1876. Fort Tejon, a single individual in August.

Contra Costa County. W. E. Bryant.—Two specimens taken in March.

Central California. L. B.—Rather rare during migrations in the valleys and foothills, though an uncertain number breed on the Sacramento and Feather rivers. First seen at Stockton April 22, 1879, April 28, 1880. Dr. Cooper informs me that he saw it at Emigrant Gap, on the Central Pacific Railroad, altitude 5,911 feet, April 29, 1870. It arrived at Nicasio April 5, 1876, according to Mr. C. A. Allen; at Beaverton, Oregon, according to Mr. Anthony, on April 17, 1885, and became common on the 30th of April, a few breeding there. It is common in the densest fir forests of the Sierras of Central California from about the first of May until the middle of September, nesting frequently in dogwood *(Cornus Nuttallii)*. I have also found nests in fir trees, on oaks, on Libocedrus, and in shrubs *(Ceanothus)*. I suppose Dr. Cooper refers to this form when he says the solitary vireo is common in Washington Territory, arriving from the south in May.

Suckley, 1860. Rather common near Fort Steilacoom.

Henshaw, 1879. The eastern slopes of the Sierra and Cascades appear to furnish, in summer at least, nothing but typical *cassinii*, while from the Calaveras Grove, California, and other localities on the western side of the mountains, we have both typical and intermediate specimens, both styles breeding.

Ridgway. Noticed only in the cañons of the west Humboldt Mountains, where it is not uncommon in September.

224. Vireo solitarius plumbeus (Coues). PLUMBEOUS VIREO.

Ridgway. The first locality where we met with this species in traveling eastward was the eastern slope of the Ruby Mountains where several other species characteristic of the Rocky Mountain district were first encountered, as *Helminthophila virginiæ*, and *Selasphorus platycercus*. It was rather common in July and August, its usual abode being the cedar and nut-pine groves on the lower slopes of the mountains, along with *Dendroica nigrescens* and *Helminthophila virginiæ*, or in the brushwood of the ravines. Certain of its notes so closely resembled those of the western house wren that they were hard to distinguish.

Fort Tejon. Henshaw, 1876. A single specimen August 1, in much worn plumage.

225. Vireo huttoni CASS. HUTTON'S VIREO.

San Diego. L. B.—Rare in winter; not seen south of San Diego.

Cooper, 1870. At San Diego I shot a female on the 9th of March, containing an egg nearly ready to be laid.

Santa Catalina Island. F. Stephens.—August, 1886, secured a female.

San Bernardino. F. Stephens.—Rare winter visitant of the foothills, possibly very rare summer resident of the same.

Henshaw, 1876. At Santa Barbara in June quite common; breeding.

Santa Cruz. Joseph Skirm.—Quite rare.

Santa Cruz. A. M. Ingersoll.—Eggs collected here; also at Olema, April 15, at which time and place I found nest and young.

Oakland and vicinity. W. E. Bryant.—Tolerably common in spring.

Berkeley. T. S. Palmer.—Rare, February 13 and March 25, 1886.

Nicasio. C. A. Allen.

Sebastopol. F. H. Holmes.

Calaveras and Yuba counties. L. B.—Rare in winter; Colfax, November 18, rare. At Chico, February 5, 1885, I saw Mr. William Proud shoot one which with several others appeared to be wintering in General Bidwell's garden.

226. Vireo bellii pusillus (Coues). LEAST VIREO.

Breeds rather commonly at Marysville and Stockton. First seen at San Diego, April 19, 1884; a male shot. First seen at San Diego, April 2, 1885; several. Rather common summer resident in the willows near San Diego.—L. B.

Poway. F. E. Blaisdell.—First seen April 28, 1884; May 21, nest and four eggs. First seen April 6, 1885. Last seen in 1884, October 5.

San Bernardino. F. Stephens.—Common summer resident of the valley; tolerably common summer resident of the foothills.

Agua Caliente. F. Stephens.—Two April 1; one April 6; several April 14 and 15, 1886.

Henshaw. The most abundant of its tribe about Los Angeles in June. It seems to be a counterpart of *V. bellii*.

Cooper, 1870. Rather common along the upper part of Mojave River in June, 1861, and in the following spring about April 20 they began to arrive at San Diego in considerable numbers.

Ridgway. This vireo was the characteristic and most abundant species at Sacramento City. Its notes are somewhat different from those of *V. bellii*.

[I have heard both and could not detect any difference. Their nests are usually if not always a few feet from the ground in dense thickets].—L. B.

Baird, Brewer and Ridgway, Appendix, vol. 3. "Dr. Cooper found this species near San Buenaventura as early as March 26, 1872, where it was quite common. On the 23rd of April he found a nest."

227. Vireo vicinior Coues. GRAY VIREO.

F. Stephens. Very rare summer resident of the foothills of San Bernardino County, also of Campo and

Julian, San Diego County, first appearing at Campo about March 24, 1876.

L. B.—Campo, 1884. A pair were seen here which I thought had a nest in shrubbery or small trees, but I did not find it. Mr. Stephens' Campo specimens were the first that were collected in California. Its song resembles that of Cassin's vireo but is weaker and inferior. When I first heard it at Guaymas I thought it Cassin's vireo.

228. **Mniotilta varia** (Linn.) BLACK AND WHITE WARBLER.

A male was found on south Farallon Island by Mr. Emerson. Proc. Cal. Acad., ser. 2, i, 48.

229. **Helminthophila luciæ** (Cooper). LUCY'S WARBLER.

Cooper, 1870. This beautiful little warbler arrived from the south in the vicinity of Fort Mojave, on March 25, when my attention was first struck by its peculiar notes, resembling those of some *Dendroicas*, but fainter.

230. **Helminthophila virginiæ** (Baird). VIRGINIA'S WARBLER.

Ridgway. First observed among the cedar and piñon groves on the eastern slope of the Ruby Mountains. It was not met with west of this locality, but eastward it occurred on all those ranges having a similar or equally extensive growth. In the Ruby Mountains it was rather common in July and August.

231. **Helminthophila ruficapilla gutturalis** Ridgw. CALAVERAS WARBLER.

A specimen was collected at Fort Tejon by Xantus. At Murphys, from April 11, 1877, to about May 1, it was tolerably common, and it is not rare in the Sierra of

Central California in summer though far from numerous. Two nests found at Big Trees were on hillsides in excavations or depressions in the ground, well concealed; four eggs in one nest, five in the other. It was last seen at Summit, September 25, 1885, cold weather having driven it south.—L. B.

Agua Caliente, San Diego County, Cal. F. Stephens.—One seen on April 9 and 12, 1886.

British Columbia. John Fannin.—Rare summer resident.

Ridgway. Although not observed in summer this bird was more or less common in September in the thickets along the streams in the lower portions of the cañons.

East Humboldt Mountains, August 5, 1868, juvenile specimen.

232. **Helminthophila celata** (Say). ORANGE-CROWNED WARBLER.

Ridgway. Male and female adult, Dearing's Creek, Upper Humboldt Valley, September 11, 1868.

Camp Harney. Bendire.—Common during the migrations; a few may remain to breed.

233. **Helminthophila celata lutescens** Ridgw. LUTESCENT WARBLER.

San Diego. L. B.—Tolerably common in winter.

San Bernardino. F. Stephens.—Tolerably common migrant in valleys and foothills.

Agua Caliente. F. Stephens.—March 25–28, one shot, others seen; migrants? One seen on April 1; another April 2; every day of April 6–15, 1886.

Henshaw, 1876. Southern California; a common species in summer; Santa Cruz Island, quite numerous, breeding in June.

Santa Cruz. Joseph Skirm.—A common summer resident on the San Lorenzo River, where they nest in wild blackberry bushes. At Olema they nest on the ground.

A. M. Ingersoll.—Eggs collected by me at Santa Cruz and Olema.

Oakland. W. E. Bryant.—Tolerably common summer resident. March 7, 1885, four male specimens.

Berkeley. T. S. Palmer.—Tolerably common resident.

Nicasio. Charles A. Allen.—First seen March 13.

Seattle, W. T. O. B. Johnson.—May 1, nest and fresh eggs.

British Columbia. John Fannin.—East of Cascades; rare.

Alameda Springs. A. M. Ingersoll.—Thirteen sets of eggs found here by myself alone.

Haywards. W. O. Emerson.—March 21, 1885, one male, first seen; rare; breeds.

Sebastopol. F. H. Holmes.—March 9, 1885, first seen; next seen March 12; March 25, common. Abundant; breeds.

Beaverton. A. W. Anthony.—First seen March 19, 1885; next seen March 31, April 10; common, breeds.

Summit. L. B.—Not seen after snow storm of September 25, 1885. At Big Trees, September 20, 1880, it was still present. It is a rather common summer resident of the pine forests of Central California, and a few winter in the Sacramento Valley as far north as Marysville.

Henshaw, 1879. Moderately common in summer when inhabiting the mountain slopes, but most numerous during the fall migration.

Ridgway. The brightly colored specimens representing var. *lutescens* were prevalent in the western depression of the Great Basin, but were not observed eastward of the upper portion of the valley of the Humboldt, nor at any time during the summer.

234. Dendroica æstiva (Gmel.) YELLOW WARBLER.

San Diego. April 20, 1884, first male; April 26, first female. April 5, 1885, first male. Tehachapi, April 5, 1889, first male. Gridley, September 23, one male; unusually late. Stockton, September 14, 1881, one male; very rare in September. Common throughout California in summer.—L. B.

Poway. F. E. Blaisdell.—April 8, 1884, first.

San Jose. A. L. Parkhurst—First arrival, April 9, 1885.

San Bernardino. F. Stephens. Common summer resident.

Haywards. W. O. Emerson.—March 28, 1885, first; common April 18, 1885; common summer resident.

Henshaw, 1876. Common about Los Angeles in June.

Stockton. A. C. Davenport.—First seen April 19, 1884. (First arrival in 1885, April 8; common, April 18.—J. J. Snyder.)

Alameda and Contra Costa counties. W. E. Bryant. Tolerably common summer resident.

Berkeley. T. S. Palmer.—First, May 8, 1885; common, May 9. In 1886, first arrival April 3; common summer resident.

Marysville. W. F. Peacock.—First male April 20, 1884; bulk arrived April 21. First arrival April 3, 1885; Common April 12.

Sebastopol. F. H. Holmes.—May 10, 1885; common May 13.

Chico. William Proud.—First seen April 20, 1884.

Beaverton, Or. A. W. Anthony.—First May 5, 1885; common May 22.

Willamette Valley. O. B. Johnson.—A very common summer resident.

Walla Walla, W. T. Dr. Williams.—First arrival May 1, 1885; common May 20.

Cooper, 1860. Abundant in this Territory.

British Columbia. John Fannin. Common summer resident.

Henshaw, 1879. Numerous as a summer resident; in the shrubbery of the streams. It penetrates beyond the Columbia and into Washington Territory.

Camp Harney. Bendire.—A very common summer resident. It commences nesting about June 1.

Ridgway. Was met with everywhere in all wooded localities, with the exception of the higher forests. The most abundant and generally distributed member of the family.

Hoffman. A summer resident in most favorable localities. None were observed south of Belmont after July 5.

235. **Dendroica cærulescens** (Gmel.) BLACK-THROATED BLUE WARBLER.

Farallon Islands. Emerson.—Proc. Cal. Acad., ser. 2, i, 48.

236. **Dendroica coronata** (Linn.) MYRTLE WARBLER.

British Columbia. John Fannin.—Summer resident; not common.

Cooper, 1860. I saw on Whidby's Island in April two of the yellow-crowned warbler. I shot one. As these are the only ones I met with, it must be rare in the Territory.

Willamette Valley. O. B. Johnson.—I have obtained several birds in spring that I have referred to this species.

Nicasio. C. A. Allen.—April 17, 1876. March 15, 1884, first seen, a flock of six or eight.

Alameda and Contra Costa counties. W. E. Bryant. Rare winter visitant.

Haywards. Cooper.—I shot a very perfect male April 10, 1875 (Proc. Cal. Acad. vi, 193).

Murphys, December, 1877, common; Marysville, February, 1878, less common; Summit, October 1, 1885, first specimen. In fall and winter it is not easily distinguished from Audubon's Warbler, and being much less numerous than the latter is likely to be overlooked in California. I shot a male in fine breeding plumage at Stockton March 20, 1886. It had a companion, probably a female, migrating northward.

Burrard Inlet. John Fannin.—First seen April 17, 1885; next seen April 20; last seen May 1. It is rare, and does not breed here. *D. coronata* and *D. auduboni* come together and generally leave together.

237. **Dendroica auduboni** (Towns.) AUDUBON'S WARBLER.

San Diego. L. B.—Common winter visitant; rather common until April 9; both sexes present April 12; last seen April 18, a fine male adult.

Poway. F. E. Blaisdell.—Common winter visitant; arrived about October 1, 1883; have known it to arrive as early as September 3.

Volcan Cañon. W. O. Emerson.—February 22.

San Bernardino Mountains. F. Stephens.—Rare summer resident; abundant in the valley in winter. Agua Caliente, March 25–28, several seen, probably wintering. Seen March 19 to April 15, 1886.

L. B.—Tolerably common in Calaveras and Alpine counties in breeding season; several nests found in conifers about thirty feet from the ground, two of them were on the ends of long horizontal limbs of *Pinus contorta*, much exposed. It breeds here from 4,000 feet upward. Common in valley and foothills in winter as far north as Red Bluff.

Alameda and Contra Costa counties. W. E. Bryant. Common winter visitant.

Chico. William Proud.—Winter visitant; last seen April 25.

Beaverton. A. W. Anthony.—First seen March 22; rare until April 15; by April 21, the bulk had arrived. Most of them had gone April 25, though a few remained to breed.

Willamette Valley. O. B. Johnson, 1880.—The most abundant warbler during summer, and a few remaining until far into, if not all winter.

Cooper, 1860. The most abundant species in the territory; Straits of Juan De Fuca, March 17.

Seattle. O. B. Johnson.—March 16.

Beaverton. A. W. Anthony.—First seen March 9, 1885, next seen March 10; common March 28.

Burrard Inlet. John Fannin.—First seen April 17, 1885; next seen April 20; last seen May 10. Common summer resident of British Columbia.

Henshaw, 1879. Common summer resident at high elevations; most numerous during the migrations.

Camp Harney. Bendire.—An abundant species during the spring migrations; arrives here about May 1. None seen in the fall. A few remain to breed in the Blue Mountains.

Hoffman. In May, June and July numbers of these birds were noticed in the more fertile and timbered tributaries of the Humboldt River, though none were found south of Mt. Nagle.

Ridgway. Its migrations seem to be mainly if not entirely vertical. Specimens at Truckee Valley, December 7; Carson, April 4; West Humboldt Mountains, October 8.

Gridley. L. B.—September 20, 1884, first; September 23, many. Stockton, September 18, 1878, first.

Berkeley. T. S. Palmer.—October 2, 1885, first, one; October 7, next seen; common October 18. Abundant; does not breed. Last seen April 13, 1886.

Haywards. W. O. Emerson.—September 26, 1884, common at once; last seen April 21; common. It does not breed here.

Haywards. Dr. Cooper.—Winter visitant. September 20, 1875. Last seen April 17.

Santa Cruz. Dr. Cooper.—September 25, 1865. Last seen April 15.

Santa Cruz. A. M. Ingersoll.—October 1, 1885, a flock.

Poway. F. E. Blaisdell.—First seen September 28, 1884, and every day afterward; very common winter visitant.

238. **Dendroica maculosa** (Gmel.) MAGNOLIA WARBLER.

British Columbia. John Fannin. Common summer resident.

239. **Dendroica nigrescens** (Towns.) BLACK-THROATED GRAY WARBLER.

San Diego, April 20, 1884, first seen. Rare here, and only during the migrations. First, May 3, 1885, male. Tehachapi, April 5, 1889, first male.—L. B.

Julian. N. S. Goss.—May 4.

San Bernardino. F. Stephens.—Rare summer resident of the mountains; tolerably common migrant in the mountain valleys and foothills. Agua Caliente, one, April 8; two, April 12 and 13; three, April 14, 1886.

Henshaw, 1876. Common in the mountains near Fort Tejon in early August, and I think they find here, in the pine region, their summer haunts.

Central California. L. B.—A rare migrant through the valleys, less so in the foothills and mountains. A few breeding in the pine region. It was first seen at Stockton April 14, 1879; April 25, 1880. My latest fall record in the central portion of California is October 8.

Beaverton. A. W. Anthony.—May 10, 1885, first; common May 24.

Willamette Valley. O. B. Johnson.—Moderately common during summer in favorable situations, seeming to prefer dense undergrowth near a swamp. I took a nest of this species June 17, 1879, in the top of a clump of *Spirea*.

Fort Steilacoom. Suckley, 1860. Moderately abundant; generally found on oak trees.

British Columbia. John Fannin.—Rare summer resident. Arrived at Burrard Inlet May 10, 1885; common May 20.

Bendire. Observed on several occasions near the summit of the Cañon City Mountain during the early part of summer.

Ridgway. On the eastern slope of the Ruby Mountains it was abundant in July and August in the piñon and cedar woods.

Sebastopol. F. H. Holmes.—October 13, 1885.

240. Dendroica townsendi (NUTT.) TOWNSEND'S WARBLER.

San Diego. Rare migrant; first seen April 20, 1884—a male. May 3, 1885, first.—L. B.

San Bernardino. F. Stephens.—Rare transient visitant to the valleys and foothills.

Julian. N. S. Goss.—April 17, 1884.

Stockton. L. B.—April 14, 1879, first males; tolerably common and silent during the next thirty days. Stockton, April 27, 1880, first males; common until May 13; rarely seen in the pine forests of Central California. I shot one, latitude 39°, altitude 6,000 feet, September, 1878, the only time I have seen it so high in the Sierra. I have never seen it in California in winter.

Haywards. Cooper.—This morning (February 12, 1884), I saw the first *D. townsendi* ever noticed at Haywards in winter, although I am told they winter in the redwoods.

Santa Cruz. Mr. William A. Cooper.—My first specimen was taken November 3, 1878. On November 14, I shot eight specimens, not having been out since. (Bull. N. O. C., April, 1879, p. 117.) W. O. Emerson, September 12, 1886, a specimen shot on the summit between Los Gatos and Santa Cruz.

Campo. F. Stephens.—April 27, 1877, present in small numbers.

L. B.—Tia Juana, near boundary line, April 30, 1885, and May 2, three or four associated with as mamy *D. occidentalis*, moving north by short flights. San Diego, May 3, male and female shot from shade trees in the city. Only a few seen; migrants.

Poway. F. E. Blaisdell.—One specimen April 29, 1885.

Haywards. W. O. Emerson.—April 21, 1885, one male; rare.

Sebastopol. F. H. Holmes.—Rare winter visitant; first seen October 22, 1884, last seen January 29, 1885. I find it here only in winter generally in pine trees.

Baird, Brewer and Ridgway.—The species was first met with by Mr. Townsend, October 28, 1835, on the banks of the Columbia River. Mr. Ridgway met with it in the East Humboldt Mountains where it was rather common in September, inhabiting the thickest of aspens, alders, etc.

Camp Harney. Bendire.—Rare; a specimen obtained May 11, 1875. I took a nest and three eggs which I believe belong to this species. They do not resemble any other warbler's eggs in my collection. The nest was placed in a narrow ravine amongst several small willow shoots near the main stem about four feet from the ground.

Henshaw, 1879. Met with but twice in September; migrating.

241. Dendroica occidentalis (Towns.) HERMIT WARBLER.

Tia Juana. L. B.—Three or four seen April 30, 1885, and May 2.

Campo. F. Stephens.—(Bull. N. O. C., July 1883.) April 27, 1877, "coming from the south by twos and threes and even a half dozen together."

San Bernardino. F. Stephens.—Rare migrant in the valley, foothills and mountains.

Julian. N. S. Goss.—April 25, 1884.

Henshaw, 1876. A single individual taken near the head of Tule River in October.

Cuyamaca Mountains, east of San Diego. Cooper.—During the last week of April, 1872, quite common between 1,500 and 4,000 feet altitude.

Berkeley. T. S. Palmer.—Last seen August 27, 1885.

L. B.—Rare migrant through the valleys and foothills of Central California. Rare summer resident of Calaveras, Alpine, Placer and Butte counties. Both sexes shot at Galt, Sacramento County, May 13, 1880.

Suckley. I obtained two specimens in June, 1856, at Fort Steilacoom.

Burrard Inlet. John Fannin.—First seen April 20, 1885; next seen April 25; last seen May 6; rare.

Ridgway. On the 29th of August a single individual was seen in the East Humboldt Mountains.

242. Seimrus noveboracensis notabilis (Grinn.) GRINNELL'S WATER-THRUSH.

A. M. Ingersoll.—Sept 25, 1885, I took a female at Santa Cruz and I have another female that was taken by Mr. J. R. Chalker who was with me hunting a few days before I shot mine. I presented the specimen to the Smithsonian Institution.

[This is the first California record, though I found it at La Paz and Todos Santos in winter.]

A. M. Ingersoll.—The water thrush sent by you for identification is the western form *Seiurus noveboracensis notabilis*, although rather smaller and yellower beneath than any of our specimens, of which we have a very meagre series. The white spots at tips of outer tail-feathers appear to be a mere individual variation.

[Signed,] R. RIDGWAY.
November 23, 1885.

243. Geothlypis macgillivrayi (Aud.) MACGILLIVRAY'S WARBLER.

Tia Juana River, near San Diego, April 20, 1884, a male specimen, the first of the season and the last.— L. B.

Poway. F. E. Blaisdell.—First seen April 4, 1884; not seen after the middle of May. First seen March 29, 1885; tolerable common April 10, 1885.

San Bernardino. F. Stephens.—Very rare migrant in the foothills. Agua Caliente; two April 13, 1886.

Henshaw, 1876. Not detected by us from which I infer its general rarity in the southern part of the State.

San Jose. A. L. Parkhurst.—First seen April 25— five or six; rare.

Berkeley. T. S. Palmer.—Last seen April 29, 1885. First seen May 31, 1886.

Sebastopol. F. H. Holmes.—First seen April 27, 1885. Again April 30.

Haywards. W. O. Emerson. First, April 12, 1885.

Nicasio. C. A. Allen.—First seen April 8; April 20, in 1876.

Oakland and vicinity. W. E. Bryant.—Rare summer resident; a pair, of which I shot one, was breeding June 10.

Central California. L. B.—Rather rare summer resident of the Sierra. Rarely seen and only during migrations in the valley. It is moderately common when migrating in the mountains; not seen after September 25, 1885, at Summit, at Big Trees, September 20, 1880. A nest found at Big Trees June 16, 1879, was in a small *Libocedrus*, about a foot from the ground, composed of grass stalks, lined with shreds of bark of *Libocedrus* and soap root fibers; eggs, five, nearly fresh; ground-color white, spotted with brown and reddish brown. Another nest was in *Ceanothus cordulatus*.

Chico. William Proud.—First seen April 27; both sexes April 29; bulk arrived April 30, on which date I saw the first dragon-fly.

Willamette Valley. O. B. Johnson.—Summer resident, nesting quite commonly.

Beaverton. A. W. Anthony.—Common summer resident. First, May 18, 1885; common May 25.

Cooper, 1860. Very common about Puget Sound.

British Columbia. John Fannin.—Common summer resident. Arrived at Burrard Inlet, June 2, 1885; common June 16.

Henshaw, 1879. Fairly numerous in summer about the eastern slope.

Camp Harney. Bendire.—Common summer resident.

Arrives here about May 1. Nest and four eggs taken June 15.

Hoffman. General throughout the northern half of the State where the country is favorable.

Ridgway. Found in all the fertile cañons from the Sierra Nevada to the Uintahs.

Mojave River. Cooper, 1870.—I noticed the first of this species April 24.

244. Geothlypis trichas occidentalis Brewst. WESTERN YELLOW-THROAT.

San Diego. L. B.—Rare in winter in the few suitable localities.

San Bernardino. F. Stephens.—Rare resident of the valley; common in summer. Agua Caliente, March 25–28, common; perhaps resident. Seen almost every day from March 18 to April 15, 1886.

Henshaw, 1876. Apparently not very common, though distributed pretty evenly over the southern portion of the State.

Sebastopol. F. H. Holmes. — March 24; common April 12, 1885.

Alameda and Contra Costa counties. W. E. Bryant.—Tolerably common; breeds.

Central California. L. B.—Very common in tule marshes; a few winter as far north as Marysville, in the Sacramento Valley.

Beaverton, Oregon. A. W. Anthony.—First, March 21, 1885; next seen April 1; common April 5.

Willamette Valley. O. B. Johnson.—Very common resident during summer.

Walla Walla, W. T. Dr. Williams.—April 21, 1885, two; April 30, common.

Cooper, 1860. Very common in the Territory during summer. I observed its arrival about the first of April.

British Columbia. John Fannin.—Common summer resident.

Henshaw, 1879. By no means uncommon in western Nevada, in situations similar to those frequented by the species in the east.

Ridgway. In all bushy places contiguous to water this little bird was invariably to be found. Arrived at Truckee Meadows, May 10, 1868.

245. **Icteria virens longicauda** (Lawr.) LONG-TAILED CHAT.

San Diego. L. B.—April 19, 1884, male, first seen; tolerably common in mountain cañons from San Diego to Campo as late as May 15. The male was first seen April 5, 1885, at San Diego.

San Bernardino. F. Stephens.—Tolerably common summer resident of the valley.

Marysville. W. F. Peacock.—First, May 19, 1885.

Santa Cruz. A. M. Ingersoll.—Summer resident.

San Jose. A. L. Parkhurst.—First seen April 29, 1884; several, in song.

Alameda and Contra Costa counties. W. E. Bryant.—Summer resident.

Haywards. W. O. Emerson.—First, April 14, 1855, male.

Central California. L. B.—Common summer resident of dense thickets below 3,000 feet altitude; usually, if not always, near water; much oftener heard than seen. Stockton, April 27, 1879, first males seen. April 27, 1889, several males seen and heard.

Sebastopol. F. H. Holmes.—First, April 29, 1885.

Chico. Wm. Proud.—April 22, 1884, heard at 6 o'clock A. M. May 7, a specimen. April 21, 1885, first.

Wilbur, Oregon. W. E. Bryant.—Breeds.

Willamette Valley. O. B. Johnson.—During summer.

Henshaw, 1879. Extreme northern California, quite rare.

Camp Harney. Bendire. — Rare summer resident; arrives about May 15.

Hoffman. It breeds in the upper portions of Nevada.

Ridgway. Equally common in California and the interior. It arrived at Truckee Meadows May 10, 1868. Specimen at West Humboldt Mountains September 7.

Cooper, 1870. Many arrived at Fort Mojave about April 20, 1861, where on May 19 I found a nest and three eggs, besides one of the cowbird *(Molothrus)*.

246. Sylvania pusilla (Wils.) WILSON'S WARBLER.

Ridgway. Specimens taken in West Humboldt Mountains September 20, 1867; East Humboldt Mountains, August 29, and September 1, 1868.

247. Sylvania pusilla pileolata (Pall.) PILEOLATED WARBLER.

San Diego. L. B.—April 14, 1884, first males; these were common April 19.

Poway. F. E. Blaisdell.—Summer resident. First seen March 6, 1884; very plentiful on the oaks when in blossom. First seen March 20, 1885.

Julian. N. S. Goss—Breeding; seen at Los Angeles, March 6, 1884.

San Bernardino Valley. F. Stephens.—Rare summer resident. Agua Caliente. Several seen, perhaps migrants. March 19, 1886, one; seen almost every day from April 1 to 15.

Henshaw, 1876. Los Angeles. Middle of June, not uncommon in the swampy thickets. About the middle of August [at Fort Tejon ?], they became common, moving southward. The bulk of these are the true *M. pusillus*.

Santa Cruz. Joseph Skirm.—Quite common summer resident.

San Jose. First seen March 24, 1884; several.

San Jose. A. L. Parkhurst.—First seen March 22, 1885; about ten.

Olema. A. M. Ingersoll.—First seen April 3, 1884; common after April 7; breeds here and at Santa Cruz.

Haywards. W. O. Emerson.—First seen March 2; common April 18, 1885.

Berkeley. T. S. Palmer.—Last pair seen May 18, 1885.

Oakland and vicinity. W. E. Bryant.—Rare summer resident.

Sebastopol. F. H. Holmes.—First arrival April 4, 1885; common April 12.

Nicasio. C. A. Allen.—First seen March 24, 1884. April 13, 1876, first.

Beaverton. A. W. Anthony.—Common summer resident. First seen May 10, 1884; the bulk arrived May 20. First seen May 8, 1885.

Willamette Valley. O. B. Johnson.—Only noticed during spring migrations.

Suckley, 1860. Very abundant in the neighborhood of Fort Steilacoom.

Cooper, 1860. Seen two or three times, only in spring and fall.

British Columbia. John Fannin.—Summer resident. (Arrived at Burrard Inlet, April 3, 1885. Next seen May 11; common May 15).

Henshaw, 1879. As a summer resident of the eastern slope appears to be rare.

Ridgway. Not seen at Sacramento, but in the valley of the Truckee and in many suitable localities to the eastward it was a rare summer resident, becoming exceedingly numerous in autumn.

Summit. L. B.—September 25, 1885, last seen; a few no doubt breed here and in Alpine County; very rare summer resident of Calaveras County in the fir forest. Common throughout Central California during migrations. Not known to breed in the San Joaquin or Sacramento valleys.

Mr. Henshaw's east slope bird may have been true *pusillus*, as he so named it either intentionally or otherwise.

248. Setophaga ruticilla (Linn.) AMERICAN REDSTART.

Haywards. W. O. Emerson.—June 20, 1881, I shot a male. It was at rest on wild blackberry vines which ran over a low growth of willows on a creek flat.

L. B.—I have no doubt that I saw a fine male in Marysville Buttes, June 6, 1884. I was on a cliff looking for the author of a strange song when it flew past and below me exposing its distinguishing wing and tail marks.

249. Anthus pensilvanicus (Lath.) AMERICAN PIPIT.

San Diego. L. B.—Common until April 23, 1884, when it disappeared. Stockton, April 27, 1879, still here with *Zonotrichia leucophrys gambeli*, *Z. l. intermedia*, and other northern species; April 27, 1880, *Junco hyemalis oregonus*, *Z. l. gambeli*, *Z. l. intermedia* and *Z. coronata* here in considerable force; weather warm. May 2, 1880, a few *Zonotrichiæ*, etc., still remain, but the most of them have gone northward to breed. September 18, 1878, first arrival from the north. Gridley, September 24, 1884, first arrival from the north. Summit, September, 1885, first seen on the tenth. This species is very common in the agricultural districts of California in winter.

Sebastopol. F. H. Holmes.—September 27, 1884, first arrival. Abundant in winter.

Haywards. W. O. Emerson.—September 26, 1884, first arrival, three birds.

Alameda and Contra Costa counties. W. E. Bryant.—Abundant winter visitant.

Beaverton, Oregon. A. W. Anthony.—It was first seen here April 17, 1885, large flocks; April 30 it was last seen.

Willamette Valley. O. B. Johnson, 1880. Common during winter.

Cooper, 1860. Abundant on the prairies of the Territory in winter.

British Columbia. John Fannin.—East of the Cascades; accidental west.

Bendire.—Very abundant on the flats bordering Malheur Lake during the migrations and in large flocks.

Hoffman. In moderate numbers at Rose's Ranch, north of Battle Mountain and in the vicinity of Tuscarora during the latter part of May.

Ridgway. Perhaps no bird of the interior is more abundant in winter.

250. **Cinclus mexicanus** Swains. AMERICAN DIPPER.

San Bernardino Mountains. F. Stephens. — Rare resident. [Probably it never occurs in San Diego County or Lower California, owing to the absence of suitable streams.]

Dr. Cooper, 1870.—Coast Range Mountains of Santa Clara County. [Coast range of Monterey County.]

Ukiah, Mendocino County. Geo. E. Aull. — Rare resident.

Sierra of Central California. L. B.—Tolerably common resident in summer; probably not in winter, as I could not find it at Big Trees from January 6th to the 13th, 1879, when the streams were mostly covered with ice, which must have prevented it from getting its usual aquatic food. Nulato, Alaska, common, breeds and is a winter resident along the open streams: Lucien M. Turner, Signal Service Report, 1886. On the Upper Yukon, resident; E. W. Nelson, Signal Service Reports, 1887.

Willamette Valley. O. B. Johnson.—On all the dashing streams in the valley.

British Columbia.—John Fannin.—An abundant resident.

Henshaw. Common upon many of the streams of the eastern slope of the Sierra Nevadas from Carson to the Columbia River.

Sierra Valley. L. B.—June, 1885, rare.

Ridgway. Most frequently seen on the Sierra Nevada and among the western ranges of the Rocky Mountain system as the Wahsatch and Uintahs, being rarely observed in the intermediate area of the Great Basin, although it was encountered at intervals on the higher of the intervening ranges. Truckee River, November 19, a specimen.

Hoffman. I have met with it only on the western slope of the Sierra Nevada, in a cañon leading down to King's River. The cañons leading down to the eastern slope of the mountains toward Independence were also well watered, but no specimens were noted, although they may occur.

251. **Oroscoptes montanus** (Towns.) SAGE THRASHER.

San Diego. L. B.—May, 1881, two specimens; apparently rare in San Diego County and southward, although Dr. Heermann " remarked it on several occasions in the environs of San Diego and from thence to Fort Yuma " (Vol. x, P. R. R. R.). Mr. Godfrey Holterhoff wrote May 30, 1884, I have not yet seen it. It was common last summer near National City, four miles from San Diego.

San Bernardino. F. Stephens.—Tolerably common migrant in foothills.

Ventura County. Evermann.—Rare migrant; one specimen March 11, 1881.

Burrard Inlet, British Columbia. John Fannin.—April 15, 1885, I saw one mountain mockingbird, an unusual occurrence here. The bird lit on a limb about ten feet from me and remained about a minute.

Baird, Brewer and Ridgway. Nuttall met with it nearly as far north as Walla Walla.

Henshaw, 1879. It is nowhere more abundant than on the sage covered hills and plains of western Nevada just at the base of the mountains which shut off the westward extension of the species.

Camp Harney. Bendire.—Common summer resident, one of the earliest birds to arrive in spring. It nests in various bushes, principally sage and serviceberry bushes. I believe that two broods are raised in a season. They leave here about the middle of September.

Ridgway. It is distributed entirely across Nevada, arrives at Carson from the south about March 20. The males begin to sing about March 24. It leaves the latitude of Carson in October or November.

Sierra Valley, Sierra County, Cal. L. B.—Common in sagebrush, June 1885. This locality can hardly be considered on the east slope, although it has its main characteristics and several of its species.

Hoffman. Of frequent occurrence throughout the northern and middle sections of Nevada, usually in greater numbers in the vicinity of streams. None were seen south of Belmont though favorable localities were found.

252. Mimus polyglottos (Linn.) MOCKINGBIRD.

San Diego. L. B.—Tolerably common resident; rare above 1,500 feet altitude.

Santa Catalina Island. F. Stephens.—August, 1886, common.

Poway. F. E. Blaisdell.—Not very common; constant resident.

Julian. N. S. Goss.—One bird, May 13, 1884.

San Bernardino. F. Stephens.—Common resident; breeds in the valleys. Agua Caliente, several seen—probably resident.

Santa Ana Plains, Los Angeles County. F. E. Blaisdell.—December 10-14, common.

Henshaw, 1876. Along our route from Los Angeles to Santa Barbara it was seen on a few occasions only.

Cooper, 1870. Salinas Valley near Monterey.

L. B.—It formerly nested in Marysville and in Marysville Buttes, and perhaps still does so. I noticed one at Gridley, July 22, 1885. In the foothills of San Joaquin and Calaveras counties it is a rare resident. December 17, 1879, I saw four in Captain Meaders' garden at Copperopolis, feeding on berries of Californian holly (*Heteromeles arbutifolia*). I have seen a few at different times here and about twenty miles northward when shooting in winter; in fact, rarely hunt here two or three days without seeing one or more in winter or summer.

Chico. William Proud.—February 10, 1884, one seen. It remained several days.

253. Galeoscoptes carolinensis (Linn.) CATBIRD.

Farallon Islands, near San Francisco. Charles H. Townsend (Auk, ii, 215).—September 14, 1884, one specimen.

Fort Walla Walla, Washington Territory. J. W. Williams.—June 2, 1885, four birds; again seen June 10, but not afterward; not seen here before.

British Columbia. John Fannin.—Rare summer resident.

254. Harporhynchus redivivus (Gamb.) CALIFORNIAN THRASHER.

San Diego. B. F. Goss.—March 16, 1884, two nests and full sets of eggs.

San Diego. L. B.—April 12, 1884, two broods of young just out of nests; moderately common here.

Poway. F. E. Blaisdell.—Common resident.

San Bernardino. F. Stephens.—Tolerably common resident of the valleys and foothills.

Oakland. W. E. Bryant.—Rare winter visitant.

Berkeley. T. S. Palmer.—Rare accidental visitant.

Ukiah.—G. E. Aull.—Tolerably common resident.

Central California. L. B.—Very common in the foothills, occasionally seen in the valley thickets.

Red Bluff. February 3–5, 1885, quite common and no doubt constant resident, as the species is not migratory to any extent. Mr. B. W. Evermann informed me February 16, 1881, that it was then nesting at Santa Paula, Ventura County. It was mated, and probably doing the same in the lower part of Calaveras County a week later. The spring of 1881 was unusually early, that of 1884 unusually backward.

255. Harporhynchus lecontei (Lawr.) LECONTE'S THRASHER.

Gila River, Fort Yuma, Mojave River. Baird, Brewer and Ridgway.

F. Stephens (Auk, October, 1884.) Extreme western end of the Colorado Desert at the foot of the San Jacinto Peak, rare; nest and eggs taken.

Agua Caliente, San Diego County. F. Stephens.—March 22, 1886, one; several seen between March 29 and April 15.

Godfrey Holterhoff, Jr. (Bull. Nutt. Orn. Club, January, 1883.) Flowing Wells, about seventy-five miles north of Fort Yuma, nest and eggs in a Palo Verde tree.

Cooper, 1870. Rather common on the deserts along the route between the Colorado Valley and the coast slope of California. They were so very wild I could obtain but two.

256. Harporhynchus crissalis (Henry). CRISSAL THRASHER.

Cooper, 1870. Rather common at Fort Mojave.

257. Campylorhynchus brunneicapillus (Lafr.) CACTUS WREN.

San Diego. L. B.—Rather common resident; rarely seen in winter. First heard January 19; began to be very noisy February 8, in spring-like, sunny weather. April 3, nest just finished; probably intergrades with *C. affinis* of the Cape region.

Poway. F. E. Blaisdell.—Tolerably common resident. Very common in Santa Ana Plains, Los Angeles County, December 10 to 14.

San Bernardino. Common resident in deserts and desert-like parts of the valleys where it breeds.

Agua Caliente. F. Stephens.—Not common resident; set of eggs taken March 28.

Ventura County. Evermann.—A common summer resident where *cacti* are abundant.

Kernville. Henshaw, 1876.—One or two individuals shot.

Hoffman. Met with only in the sandy deserts about 30 miles northwest of Fort Mojave among the cactus and Yucca.

258. Salpinctes obsoletus (Say). ROCK WREN.

San Diego. L. B.—Rare resident.

Poway. F. E. Blaisdell.—Resident. Santa Ana Plains, December 10–14, frequently seen and heard.

Henshaw, 1876. Island of Santa Cruz.

San Bernardino. F. Stephens.—Rare resident; tolerably common transient visitant to the valley and foothills.

Agua Caliente. F. Stephens.—Foothills, not common. One seen April 12, 1886.

Tehachapi. L. B.—Unusually common, apparently constant resident.

Alameda and Contra Costa counties. W. E. Bryant.—Rare resident.

Berkeley. T. S. Palmer.—Tolerably common resident.

Nicasio, Marin County. Charles A. Allen.

Sebastopol. F. H. Holmes.

Central California. L. B.—Stockton, one in a brickyard in winter, a migrant; foothills near Copperopolis, Murphys, Marysville Buttes, Oroville, Colfax, Summit in summer up to altitude 8,500 feet. Resident, but never numerous below altitude 2,500 feet. More common in summer about Summit than any place where I have seen it.

Farallon Islands. Messrs. W. O. Emerson and A. M. Ingersoll were on these islands in June, 1885, and found a few old and young birds.

Beaverton, Oregon. A. W. Anthony.—May 21, 1885, one shot; the only one seen.

British Columbia. John Fannin.—Common summer resident.

Henshaw, 1879. A common summer visitant throughout this whole region.

Camp Harney. Bendire.—Common summer resident;

one of the earliest birds in spring and one of the latest in fall.

Ridgway. Found in suitable localities in all the desert ranges. It arrived at Carson, March 20, 1868.

259. Catherpes mexicanus conspersus Ridgw. CAÑON WREN.

San Diego. L. B.—Rare; always found, however, in Mission Cañon.

Poway. F. E. Blaisdell.—Seen several times in February and March.

San Bernardino. F. Stephens.—A rare transient visitant to the foothills.

Henshaw, 1876. In the neighborhood of Mt. Whitney tolerably numerous. It was detected, too, at various points in the Coast Range, so that its diffusion over southern California may be said to be general.

Mt. Diablo. W. E. Bryant.—A pair seen.

L. B.—Foothills of Central California in suitable localities north to Oroville and probably to near base of Mt. Shasta; Murphys, Colfax and Oroville in winter; breeds on the Stanislaus River, altitude 4,000 feet. Summit, Central Pacific Railroad, October 4, 1885, a single specimen journeying westward from the east slope; nowhere numerous.

Ridgway. We found it everywhere more rare than the rock wren and apparently confined to the more secluded portions of the mountains. An adult male specimen, near Fort Churchill, on the Carson River, December 7, 1867.

Chico. William Proud—June 5, 1885, up under the bluffs, quite a colony, and with them was *Salpinctes obsoletus*. I was botanizing, had no gun and took no specimens, but I am certain of the species. They were very tame so that I was within ten feet of them. They were running and creeping through every little crack and fissure in the rocks.

Shasta County. Charles H. Townsend.—Often seen in the lime rocks at Baird; young observed late in June. But one was found on the lava rocks above the timber line of Mt. Shasta.

260. **Thryothorus bewickii spilurus** (Vig.) Vigors's Wren.

San Diego. L. B.—Rare.

Poway. F. E. Blaisdell.—Several specimens taken.

Santa Ana Plains. F. E. Blaisdell.—December 10-14 tolerably common.

San Bernardino. F. Stephens.—Common summer resident of the valley and foothills; rare in winter.

Henshaw, 1876. Common resident during summer in the southern half of California.

Santa Cruz. Joseph Skirm.—Common.

Berkeley. T. S. Palmer.—Common winter resident; rarely breeds. Last seen February 19, 1886.

Alameda and Contra Costa counties. W. E. Bryant.—Common resident.

Central California. L. B.—Common resident in suitable localities; breeding less commonly in the Sierra Nevada than in the valleys.

Ukiah. G. E. Aull.—Rare resident.

Beaverton, Oregon. A. W. Anthony.—Usually common; rare this year (1885).

Puget Sound. Suckley, 1860.—Very abundant; a constant resident throughout the year, and is not less abundant in winter.

British Columbia. John Fannin.—Common summer resident.

Camp Harney. Bendire.—Rather rare in this vicinity.

Ridgway. After leaving Sacramento we nowhere identified it with certainty.

Cooper, 1870. During the winter a few birds in the vicinity of Fort Mojave, but left probably for the mountains in April.

Newberry. Not uncommon in bushes and among fallen logs between Fort Reading and the Columbia.

[Judging by one of Mr. Anthony's Beaverton specimens the Oregon coast bird is much darker than the Californian].

261. Troglodytes aëdon parkmanii (Aud.) PARKMAN'S WREN.

San Diego. L. B.—Common resident. El Cajon, San Diego county, January 16, one specimen; Marysville, December 23, one good specimen, and on the following 'day another; probably driven out of the river bottom by flood of December 22 and 23, 1884.

Poway. F. E. Blaisdell.—Plentiful since the first of March; several nests found. Volcan Mountains, from August 28 to November 28; common.

San Jose. A. L. Parkhurst.—First, March 1, 1885; common and singing, March 15.

Volcan Mountains. W. O. Emerson.—One specimen, January 24; also seen on the 28th. (Arrived at Haywards March 6, 1885; common, March 18.)

San Bernardino. F. Stephens.—Rare transient visitant and resident; very rare in the foothills. Agua Caliente, March 25, one shot. April 12 and 13, 1886, one each day.

Henshaw, 1876. The most numerous of its tribe.

Oakland and vicinity. W. E. Bryant.—Common summer resident. March 7, 1885, one male.

Berkeley. T. S. Palmer.—Common summer resident. Arrived March 29, 1885; common April 1. In 1886, arrived March 31; common April April 4.

Sebastopol. F. H. Holmes.—First, April 13, 1885; common April 19.

Nicasio. Charles A. Allen.—Arrived April 8 (1884?). April 2, 1876.

Mountains of Central California. L. B.—Common summer resident; less common in the valleys during the same time.

A specimen sent me by Mr. Anthony from Beaverton, Oregon, was barred heavily below; the dark bands above were very distinct. The bird was much darker than the Californian examples; however, the birds near the coast, especially northward, are well known to be darker than those of the interior, and this is the basis of several varieties. March 30, 1885, first seen; next seen April 1; common April 28. (Anthony.)

Marysville. W. F. Peacock.—December 15, 1885, a specimen. (I shot one at Marysville, December 23, and another December 24, 1884.—L. B.)

Willamette Valley. O. B. Johnson, 1880.—Common during the summer.

Puget Sound. Cooper, 1860.—Common. It arrives about April 20.

British Columbia. John Fannin.—Common summer resident.

Henshaw, 1879. Numerous as a summer resident all along the eastern slope.

Camp Harney. Bendire.—A very common summer resident, abundant wherever there is any timber. It commences nesting about June 1.

Ridgway. Equally abundant among the cottonwoods of the river valleys and the aspen copses of the higher cañons; abundant in the high cañons of the East Humboldt and Wahsatch Mountains.

Hoffman. Widely distributed in Nevada; was found breeding at Morey in June. None were found south of Belmont after July 1.

262. Troglodytes hiemalis pacificus Baird. WESTERN WINTER WREN.

British Columbia. John Fannin.—Common resident.

Cooper, 1860. Probably the most common species in the Territory. Most commonly seen in winter.

O. B. Johnson, 1880. Remains during the winter, but leaves for other parts to breed.

Beaverton. A. W. Anthony.—Common winter resident.

Nicasio. Charles A. Allen.—Common every winter.

Alameda and Contra Costa counties. W. E. Bryant. Rare winter visitant.

Haywards. W. O. Emerson.—Two seen October 16; again seen December 22; rare.

Big Trees. L. B.—January 6, 1879, two specimens. It appears to be rare so far south in the Sierra in winter, though it has been collected at Fort Tejon. Dunbar's, Calaveras County, May 22, 1889, several just out of the nest, all apparently of one family. South Grove, Stanislaus County, an adult seen in a pile of logs where it probably had a nest June 1, 1889.

Ridgway. Pyramid Lake, December 25, 1867, one specimen; rare.

Saticoy. Cooper. (Auk, 1887, p. 93.) Three or more of this species remained in the willows all winter, and I preserved one. This is about its most southern range.

263. Cistothorus palustris paludicola Baird. TULE WREN.

San Diego. L. B.—Rare in the few small tule patches in winter. It is a very common resident of the tule marshes of Central California.

San Bernardino. F. Stephens.—Tolerably common summer resident of the valley.

Agua Caliente. F. Stephens.—March 25-28 common in the tules at the spring, where perhaps breeding. March 23, 1886; rather common from April 1.

Newport Sloughs, Los Angeles County. F. E. Blaisdell.—December 14 to January 6, 1885, abundant in the tules.

Henshaw, 1876. Abundant in Southern California, especially in fall.

Berkeley. T. S. Palmer. December 26, 1885. Rare. Its rarity is probably due to the lack of tule swamps.

Alameda and Contra Costa counties. W. E. Bryant.—Rare.

British Columbia. John Fannin.—Common summer resident.

Camp Harney. Bendire. An abundant summer resident; a few winter here. In all marshy localities more or less abundant.

Ridgway. Truckee and Humboldt Rivers. A specimen at Pyramid Lake, December 25, 1867.

Hoffman. The southernmost locality where this species was found was in the valley immediately north of Mt. Magruder.

264. Certhia familiaris occidentalis Ridgw. CALIFORNIA CREEPER.

Henshaw, 1876. Breeds in the mountains of southern California, where I took a young bird in the first plumage near Fort Tejon, August 2. It is not common, however, until late in the fall.

Santa Cruz. Joseph Skirm.—Rare.

L. B.—Big Trees, altitude 4,700 feet; rather common in summer, breeding here and at Blood's, altitude 7,200 feet; less common at the latter locality. A nest found at Big Trees, June 3, had very young birds in it; one found June 9 had a full set of fresh eggs; other nests found here at different times were all (including the two mentioned above) in cedar trees *(Libocedrus)* between the bark and trunk, at heights varying from three to twenty-five feet. The species has been frequently seen at different localities northward, but never found numerous. At Big Trees from January 6–13, 1879, it was quite as common as in summer. Also seen at Summit, November 13–16, 1884. It rarely visits the valleys near sea level.

Cooper, 1860. Abundant in the forests. It appears to reside constantly in the Territory.

Suckley, 1860. I have obtained several specimens in the vicinity of Fort Steilacoom.

Henshaw, 1879. A common summer inhabitant of the coniferous belt along the eastern slope.

Ridgway. In winter it was more or less common among the cottonwoods in the lower part of the valleys of the Truckee and Carson rivers, but eastward it was not again met with at any season except on the Wahsatch and Uintah Mountains.

265. **Sitta carolinensis aculeata** (Cass.) SLENDER-BILLED NUTHATCH.

Campo and Escondido in January; Santa Marguerita Cañon in April.—L. B.

Poway. F. E. Blaisdell.—Observed several times in January.

Volcan Mountains. W. O. Emerson.—Seen in every walk. Mated March 1, and seemed about to breed.

San Bernardino Mountains. F. Stephens.—Rare resident.

Henshaw, 1876. Found numerously in the pine region of both the Coast and Sierra ranges.

Cooper, 1870. I saw none even in the Coast Mountains in summer near Santa Cruz.

Contra Costa County. W. E. Bryant.—Resident.

Central California. L. B.—Not rare; generally distributed in the valley in deciduous oaks and in the coniferous forest to the summit of the Sierra. The species is present in all parts of this region in summer and winter.

Ukiah. George E. Aull.—Common resident.

Willamette Valley. O. B. Johnson, 1880.—Quite common during the summer and not rare during the winter.

Suckley, 1860. Quite abundant at Puget Sound.

British Columbia. John Fannin.—Common summer resident; accidental west of the Cascades.

East Slope. Henshaw, 1879.—A numerous and constant resident among the conifers; not so common towards the Columbia River as either of the other species.

Camp Harney. Bendire.—Moderately abundant in the Blue Mountains and resident throughout the year.

Ridgway. Observed in abundance only on the Sierra Nevada, being comparatively rare on the Wahsatch and Uintah Mountains while none were seen in the intervening region.

266. Sitta canadensis Linn. RED-BREASTED NUTHATCH.

Henshaw, 1876. It appeared to be not uncommon near Mt. Whitney in October.

Central California. L. B.—Resident of the Sierra; not rare; a very rare winter visitant or straggler to the San Joaquin and Sacramento valleys. Big Trees, January 6–15, 1879. Summit, Central Pacific Railroad, November 13, 16. Butte Creek House, July 2, and various localities in summer and winter. Breeds in the fir forests only.

Sebastopol. F. H. Holmes.—Collected here.

Willamette Valley. O. B. Johnson, 1880. Associated with *S. c. aculeata*.

Cooper, 1860. Common in the Territory, preferring the oaks and other deciduous trees, and never frequenting the interior of the dense forests. I observed this *S. c. aculeata* at 49° east of the Cascade Mountains as late as October 15.

British Columbia. John Fannin.—Common summer resident; accidental west of the Cascades.

Henshaw, 1879. From the line of the railroad to the Columbia River, and so on to the north, the red-bellied nuthatch is really a common bird and in much of this area it doubtless breeds. Towards the Columbia River it was more numerous, and upon the Des Chutes its numbers in certain localities were comparable with those of the pigmy nuthatch.

Ridgway. An inhabitant in summer of the pine woods exclusively, this species was met with at that season only in the thickest or more extensive coniferous forests such as those on the Sierra Nevada, Wahsatch, and Uintah ranges. In September we found it common in the aspen groves along the streams in the upper Humboldt Valley; later in the same month it was also common in the Clover Mountains at an altitude of 11,000 feet.

Berkeley. T. S. Palmer.—Not previously collected here.

267. Sitta pygmæa Vig. Pigmy Nuthatch.

Poway. F. E. Blaisdell.—Seen several times in January.

Julian. N. S. Goss.—From the middle of March to the middle of May not seen.

San Bernardino Mountains. F. Stephens.—Rare resident.

Santa Cruz. A. M. Ingersoll.—I saw about a dozen here and one at Olema. It is quite rare.

Henshaw, 1876. By far the most abundant of the three species seen in California. Common everywhere; where the presence of pines affords them the hunting grounds they most affect.

[Mr. Henshaw refers, I suppose, to Fort Tejon, Mt. Whitney, Kernville, and other localities where he collected in 1875].

Tehachapi. L. B.—March 30, 1889, a small flock in pine forest.

Baird, Brewer and Ridgway. Dr. Gambel mentions their almost extraordinary abundance in the winter months in upper California. Around Monterey at times the trees appeared almost alive with them.

L. B.—The type was collected at Monterey, south of which Dr. Cooper says he has not seen the species. Dr. Heermann does not appear to have met it during about three years collecting in California. Dr. Newberry says: " We saw it in most wooded places where water was near and any considerable amount of animal life was visible," probably refering to the region east of the Sierra Nevada and Cascades. I saw a few at Big Trees in July, 1878, and have not been able to find the species in California since then, except at Tehachapi in 1889, and four individuals during ten days collecting in Monterey County. Dr. Cooper says: "It was met with only in the open pine forests at Fort Colville, near the 49th degree. * * * This bird, like many other California species, probably migrates only along the east side of the mountains, shunning the damp spruce forests near the coast," all of which taken together tends to prove that it has never been generally distributed west of the mountains.

British Columbia. John Fannin.—Common summer resident; accidental west of the Cascades.

Henshaw, 1879 (east slope). The most numerous of

the family in the Sierra Nevada and Cascade Mountains, as almost everywhere through the west.

Carson. Ridgway.—Abundant winter resident, breeding on the east slope of the Sierra Nevada.

Camp Harney. Bendire.—A moderately abundant summer resident in the Blue Mountains. A few remain throughout the year.

268. Parus inornatus Gamb. PLAIN TITMOUSE.

San Diego County. L. B.—Rare resident.

Volcan Mountains. W. O. Emerson.—January 30. Mated and singing March 2.

Volcan Mountains. F. E. Blaisdell.—From August 21 to November 28 rather common.

San Bernardino. F. Stephens.—Rare winter visitant.

Berkeley. T. S. Palmer.—Last seen April 14, 1886.

Alameda and Contra Costa counties. W. E. Bryant.—Tolerably common resident.

Ukiah. G. E. Aull.—Common resident.

L. B.—Central California up to about 5,000 feet, common resident; Red Bluff, February 3, 1885, common.

Ridgway. Pine forests of the eastern slope, especially, in their lower portions, a rather common species; common at Carson in winter.

Henshaw, 1879. Present in Nevada in the foothills of the mountains and on the low ranges to the east of the main chain. It was not met with in the Columbia River region, nor even in northern California.

269. Parus inornatus griseus Ridgw. GRAY TITMOUSE.

New Mexico and Colorado to Arizona and Nevada.

270. Parus atricapillus occidentalis (Baird). OREGON CHICKADEE.

British Columbia. John Fannin.—Abundant resident.

Cooper, 1860. I observed its nest near Puget Sound.

Suckley, 1860. Quite abundant in the valley of the Willamette, also at Fort Vancouver.

Beaverton. A. W. Anthony.—Common resident.

Willamette Valley. O. B. Johnson, 1880.—Common throughout the year.

Wilbur, Oregon. W. E. Bryant.—Specimens in the breeding season of 1883.

Camp Harney. Bendire—Common during winter retiring to the mountains to breed.

271. Parus gambeli Ridgw. MOUNTAIN CHICKADEE.

Poway. F. E. Blaisdell.—February 15, several seen; one shot.

Volcan Mountains. W. O. Emerson.—February 24, and occasionally afterward; on March 1, singing and looking for nesting sites.

Volcan Mountains. F. E. Blaisdell.—From August 21 to November 28, tolerably common.

San Bernardino Mountains. F. Stephens.—Tolerably common resident.

Central California. L. B.—Common resident in conifers, Big Trees, January 6–13, 1879. Summit, Central Pacific Railroad, November 16; also at Donner Lake, November 16.

Ukiah. George E. Aull.—Tolerably common resident.

British Columbia. John Fannin.—Abundant visitant.

Henshaw, 1879. Breeds numerously among the pines; extremely abundant among the oaks of the eastern slope near the Columbia River.

Camp Harney. Bendire.—Common during the winter in the willows and shrubbery. In summer they breed on the higher mountains.

Ridgway. We found it in all pine forests as well as the more extensive of the pinon and cedar woods on the interior ranges.

272. Parus rufescens Towns. CHESTNUT-BACKED CHICK-ADEE.

British Columbia. John Fannin.—Abundant resident.

Cooper, 1860. The most abundant species in the forests of this Territory. It appears to prefer the evergreens, where large parties of them may be found at all seasons.

Willamette Valley. O. B. Johnson.—Less abundant than *P. occidentalis* which they closely resemble in habits.

Beaverton. A. W. Anthony.—Common resident; nest found April 28.

Wilbur, Oregon. W. E. Bryant.—Specimens in breeding season of 1883.

[Perhaps the last three belong under *P. r. neglectus.*]

273. Parus rufescens neglectus Ridgw. CALIFORNIAN CHICKADEE.

Santa Cruz. A. M. Ingersoll.—Breeds here and at Olema; have collected its eggs at both places.

Sebastopol. F. H. Holmes.—I have shot it here.

Ukiah. George E. Aull.—Rare resident.

Monterey and Santa Cruz counties. L. B.—Appears to be decidedly rare.

274. Chamæa fasciata Gamb. WREN-TIT.

San Diego. B. F. Goss.—March 16, 1884, two nests with fresh eggs, one containing four, the other five.

San Diego. L. B.—Common resident.

Poway. F. E. Blaisdell.—Common resident; nests found soon after the middle of March.

San Bernardino. F. Stephens.—Tolerably common resident of foothills.

Alameda and Contra Costa counties. W. E. Bryant.—Tolerably common resident.

Berkeley. T. S. Palmer.—Abundant resident.

Sebastopol. F. H. Holmes.—Shot here.

Red Bluff. L. B.—Common resident; also of Colfax and Murphys. Alta, November 17, 1884, not rare, though seldom found so high (3,600 feet). I suppose the Alameda and Contra Costa County notes refer to the typical bird, the others to variety *henshawi*, although specimens from Red Bluff, Oroville, Colfax, Gridley, Marysville, Stockton, Calaveras County and San Diego, are much darker than the type of *henshawi*, as I remember it, having seen it in 1882. This came from Walker's Basin, near Caliente, as Mr. Henshaw informs me. I doubt if it is really abundant at any locality, but is quite common in most parts of California, in shrubbery and thickets, where it finds shelter.

275. **Chamæa fasciata henshawi** Ridgw. PALLID WREN-TIT.

Interior of California, including the western slope of the Sierra Nevada.

276. **Psaltriparus minimus** (Towns.) BUSH-TIT.

British Columbia. John Fannin.—Common resident.

Cooper, 1860. Quite abundant during summer at Fort Steilacoom. They arrive towards the middle of April.

Beaverton. A. W. Anthony.—First seen March 11, 1885; March 12, tolerably common; breeds.

Willamette Valley. O. B. Johnson.—Plentiful during the winter months among the evergreens; always in small flocks. Many remain all summer to breed, but they are more retired and less conspicuous. (1880).

277. Psaltriparus minimus californicus Ridgw. CALIFORNIAN BUSH-TIT.

Red Bluff. L. B.—Rather common February 3, 1885, in pairs. Mr. B. W. Evermann, of Santa Paula, Ventura County, informed me that birds of this species were laying there February 16, 1881. I noticed that it was mated at the same time in the foothills of San Joaquin County. It is common and pretty generally distributed in most parts of California below the fir forests; rarely seen in them, but I have seen it at 5,000 feet altitude.

Ukiah. George E. Aull.—Common resident.

Chico. William Proud.—April 16, 1884, nest finished.

Berkeley. T. S. Palmer.—Abundant summer resident.

Alameda and Contra Costa counties. W. E. Bryant.—Common resident.

San Jose. A. L. Parkhurst.—January 15, 1885, flocks were breaking up in pairs; February 15, building.

Poway. F. E. Blaisdell.—Common resident; nest April 26, 1884.

San Diego. L. B.—Rare, owing to scarcity of trees and shrubbery, but more common in the mountains on the east and at Campo. I suppose the bird south of Oregon, especially in the interior, to be var. *californicus* Ridgw. The Red Bluff, Oroville and Colfax birds, I collected in winter, are of this form, and probably so at all times. Mr. Henshaw appears to have found an abundance, as he got twenty-four specimens at Santa Barbara, Fort Tejon and Walker's Basin, in 1875. It was common at and about Tehachapi March and April, 1889.

San Bernardino. F. Stephens.—Common; breeds in the valleys.

Volcan Mountains. W. O. Emerson. Only seen February 24 in a snow storm, associated with mountain chickadee.

278. **Psaltriparus plumbeus** Baird. LEAD-COLORED BUSH-TIT.

Ridgway. We met with this species on several occasions from the very base of the Sierra Nevada eastward to the Wahsatch Mountains but the localities where it occurred in abundance were few and remote from each other while its habits are so erratic that it was seldom met twice at one place.

Camp Harney. Bendire.—A summer visitor, not abundant. I shot several specimens of this species in November, 1874. On June 6, 1876, I saw several near the summit of Cañon City Mountain, evidently breeding.

279. **Auriparus flaviceps** (Sund.) VERDIN.

Cooper, 1870. I found numbers of this beautiful bird at Fort Mojave during the whole winter. On the 10th of March I found a pair building. On the 27th of March I found the first nest containing eggs. There were in all cases four. I noticed the nests of this bird in the *Algarrobias* that grow in a few places on the mountains west of the Colorado Valley and along Mojave River as far west as Point of Rocks.

Agua Caliente. F. Stephens.—Western extremity of Colorado Desert, one shot March 25-28; an old nest seen. Probably this is the western limit of the species. March, 1886, bird and eggs secured.

Heermann. I found their nests abundant at Fort Yuma, though from the lateness of the season few of the birds remain.

280. **Regulus satrapa olivaceus** Baird. WESTERN GOLDEN-CROWNED KINGLET.

Haywards. W. O. Emerson.—October 16, 1884, rare.

Oakland. W. E. Bryant.—Rare winter visitant; more seen this winter (1884-85), than ever before.

Berkeley. T. S. Palmer.—January 31, 1885, rare.

Alameda. A. M. Ingersoll.—January, 1885, common.

Stockton and Murphys. J. J. Snyder.—Several specimens this winter (1884–85).

Central California. L. B.—Rare, irregular winter visitant; rare but regular summer resident in the fir forest of Calaveras County, and probably at many suitable localities throughout the firs of California.

Sebastopol. F. H. Holmes.—Abundant winter visitant; last seen March 19, 1885.

Beaverton, Oregon. A. W. Anthony.—Abundant when I arrived, February 2, 1884, decreasing in numbers soon after this time; last seen March 19. The species was last seen April 10, 1885; very rare this year.

Willamette Valley. O. B. Johnson.—Common throughout the winter in flocks.

Cooper, 1860. Abundant, especially during winter, and some remain all summer as I have seen them feeding their young in August at Puget Sound.

Ridgway. A very few individuals were noticed in the cañons of the West Humboldt Mountains.

Camp Harney. Bendire.—A few specimens November 7, 1875; not common.

Burrard Inlet. John Fannin.—Spring migrant, 1885. First seen March 14; next seen March 20; common March 29; breeds; common.

Cape Beale, British Columbia. Emanuel Cox.—First seen April 23; common May 1.

281. Regulus calendula (Linn.) RUBY-CROWNED KINGLET.

San Diego. L. B.—Rare winter visitant.

Volcan Mountains. W. O. Emerson.—Rare; perhaps breeds in the firs.

San Bernardino. F. Stephens.—Tolerably common summer resident in the high mountains and equally common in the valley and foothills in winter. Agua Caliente, March 25 and 28, several seen. Several seen from March 18 to April 15, 1886.

Oakland, and vicinity. W. E. Bryant.—Common winter visitant; begins to sing in March.

Berkeley. T. S. Palmer.—Last seen April 11; began to sing March 7. Last seen on March 25, 1886.

Sebastopol. F. H. Holmes.—An abundant winter visitant; last seen April 15; the bulk departed about March 18.

Beaverton. A. W. Anthony.—First seen March 9, two birds; again seen March 12; last seen April 15. It was common during the migration.

Burrard Inlet. John Fannin.—First seen March 14; was common. April 4. Common in breeding season.

Henshaw, 1879. A common summer inhabitant of the pineries.

Camp Harney. Bendire.—Undoubtedly breeds about here. A number remain amongst the willows and alders during the winter.

Ridgway. A common winter resident in all the lower valleys, while in early spring it became abundant to such an extent as to exceed all other birds in numbers.

L. B.—Common and generally distributed in the timbered parts of the valleys and foothills of Central California in winter; common in breeding season in the upper Sierra from latitude 38° northward, and when a heavy snow-fall makes a backward spring, a few breed as low as the Calaveras Big Trees. The first migrants appeared at Gridley, October 1, 1884.

282. Polioptila cærulea obscura Ridgw. WESTERN GNAT-CATCHER.

San Diego. L. B.—Tolerably common in winter; nest and four fresh eggs at Campo, May 14, 1884; nest and one egg, Stockton, May 7, 1879; two nests at Murphys May 28, 1880, one containing four, the other five eggs; all nearly fresh.

Poway. F. E. Blaisdell.—Breeds here, Santa Ana Plains, December 10, 1884, very plentiful; Santa Ana River December 14 and later, not common.

Julian. N. S. Goss.—April 2, 1884.

San Bernardino. F. Stephens.—Common summer resident; common at Agua Caliente March 25 and 28, 1884. Almost every day from March 19 to April 15, 1886. *P. melanura* is probably the summer form at Agua Caliente.

Henshaw, 1876. Fort Tejon, particularly numerous July 27, August 8. Neither here nor elsewhere was the closely allied " *P. melanura* " detected.

Contra Costa County. W. E. Bryant.—Tolerably common summer resident.

Central California. L. B.—Rather rare summer resident up to 2,500 feet altitude. My earliest Stockton record is March 12; my latest, November 22, when I saw two birds and shot one in cold weather.

Cooper, 1870. I found them at Fort Mojave, March 20.

283. Polioptila plumbea Baird. PLUMBEOUS GNAT-CATCHER.

Habitat. Southwestern border of the United States, from southern Texas to lower Colorado Valley, and thence south to Cape St. Lucas; northern Mexico (Ridgway's Manual).

I have no other authority for giving this a place here

than that above cited, but have no doubt that it occurs in San Diego County east of the mountains, as it is very common about La Paz and other parts of the Cape region in winter, and, like other species, extends from Cape St. Lucas to the Colorado Valley and Arizona on the east side of Lower California. I have not seen the perfect black crown in winter, though I shot one at La Paz, January 12, with crown nearly black, and recorded it as it was the only one I had seen in this plumage in winter. The black-crowned *Polioptila* that breeds in the Cape region utters the same scolding, cat-like squall that the San Diego bird does, a squall I have never heard *P. cærulea* utter. The boys at San Diego call *P. californica* the cat bird.

284. Polioptila californica Brewst. BLACK-TAILED GNAT-CATCHER.

San Diego. L. B.—Probably seen in December and January, 1883–84, but I could not find it in February and March, when there were frequent chilly rains, nor did I succeed in getting a specimen until April 7, when I shot a fine male. It was rare in the spring of 1884, though common in April and May, 1881. Col. N. S. Goss, while at San Diego, informed me that he had two specimens which he shot in December, 1882, at San Diego, and that he saw the species at Los Angeles March 6, 1884. I suppose this is the species he referred to, though he named *P. plumbea* or *P. melanura*. I have no doubt many *P. californica* winter in Lower California, though I did not get an undoubted specimen in the southern part of the Peninsula. I think it breeds as far south at least as Santa Rosalia Bay.

Dr. Cooper says of *P. melanura* (Cal. Orn): " This was also a rather common bird during the whole winter at Fort Mojave, as well as at San Diego, and I obtained one in October on Catalina Island, but did not find it there or at Santa Barbara in summer."

Poway. F. E. Blaisdell.—Rare. In 1876 a pair of *Polioptilas*, the male having a black cap, had a nest in forks of a dead scrub oak, four feet from the ground. It contained five young birds. At Santa Ana Plains, December 14, I obtained one specimen.

San Bernardino. F. Stephens.—A very rare summer resident of the foothills.

Ventura County. B. W. Evermann.— Not so common as *P. cœrulea*. Resident.

[One of Mr. Brewster's type specimens was collected at Saticoy by Dr. J. G. Cooper; another was from San Bernardino, and another was from Fort Yuma.]

285. Myadestes townsendii (Aud.) Townsend's Solitaire.

San Diego. L. B.—January 24, 1884, one specimen, the only one seen by me so far south. It had been eating manzanita berries *(Arctostaphylos)*.

Colorado Desert. F. Stephens.—One seen on March 21, 24, and April 2, 1886.

Poway. F. E. Blaisdell.—February 23, 1884, a single specimen shot during a storm, the first I have seen here. I noticed the species at Temecula, November 12, 1883.

Ballena. W. O. Emerson.—Volcan Mountains, January 23, 1884, the day of my arrival, I saw a single bird, and two more a few days later.

Oakland. W. E. Bryant.—Rare winter visitant.

Central California. L. B.—West slope of the Sierra in pine forests, rather common summer resident though never numerous; begins to lay about June 1, nests usually on the ground, more or less secreted.

Ridgway. Virginia Mountains near Pyramid Lake, December 21, 1867.

Henshaw, 1879. Very abundant in the Des Chutes Basin in September, where, too, it was reported by Dr. Newberry in 1860. In fall and winter it appears to be generally dispersed over much of the country adjoining the eastern slope where in summer it appears to be almost entirely absent. In the summer of 1875, Mr. Henshaw did not see the species about Fort Tejon.

Camp Harney. Bendire.—Rather common among the juniper groves during spring and fall and in mild winters throughout the whole season. None remain during the breeding season.

Burrard Inlet. John Fannin.—May 20, 1885, only two specimens seen. It does not breed here.

Summit. L. B.—From about September 15, 1885, to October 10, very common in the junipers and moving southward by short flights. In November, from the 13th to 16th, 1884, several seen here.

286. Turdus ustulatus (Nutt.) RUSSET-BACKED THRUSH.

Common south of Campo, May 12, very common between Campo and San Diego May 16, and probably breeding. The small thrushes of the Pacific Coast are very difficult to separate without specimens in hand, and even with them mistakes of identification are likely to occur, though the present species is found in this district only in summer, and in a considerable portion of it the dwarf thrush only in winter. The songs of these species differ greatly. I requested Mr. Blaisdell to give careful attention to the arrival of this species in San Diego County in 1885, and it will be seen that the date given for 1885 is much later than in 1884, a cool spring.

Poway, San Diego County. F. E. Blaisdell.—First seen April 15, 1884. First seen May 1, 1885; last seen May 4, 1885.

W. O. Emerson.—Seen in the cañon at the foot of Volcan Mountain April 3; at Poway April 10, 1884.

Mr. Emerson did not get a specimen April 3, perhaps not on the 10, 1884.—L. B.

San Bernardino. F. Stephens.—A tolerably common summer resident of the valley.

Agua Caliente. F. Stephens.—One April 7, 1886.

Santa Barbara. Henshaw (1876).—Young fully fledged by the last of June.

Santa Cruz. Joseph Skirm.—Commenced nesting about May 15, 1884; three to four eggs, the latter the most I have found in any one of about one hundred and fifty nests.

San Jose. A. L. Parkhurst.—The first seen April 25, 1884; four individuals.

Nicasio. Charles A. Allen.—April 23, 1884, the first seen.

Olema. A. M. Ingersoll.—The first seen May 7, 1884.

Chico. William Proud.—First seen May 7, 1884; five birds.

Beaverton. A. W. Anthony.—A common summer resident; breeds here.

San Diego. L. B.—May 3, 1885, first seen; five birds in shade trees and gardens; May 5, both sexes shot; mostly gone by the 9th.

Poway. F. E. Blaisdell.—May 1, one bird, first seen; May 2, 1885, two birds; common summer resident; breeds here.

Stockton. L. B.—First seen May 10, 1880.

Haywards. W. O. Emerson.—April 12, two, first seen; common April 18, 1885; breeds. Last seen September 13, 1885.

Oakland. W. E. Bryant.—Common summer resident.

Berkeley. T. S. Palmer.—April 23, one, first seen;

next seen April 25; common April 27; breeds here; began to sing May 2, 1885. In 1886 the first seen was on April 25; common April 30; began to sing April 28.

Chico. Wm. Proud.—First seen May 1; common May 6, 1885.

Sebastopol. F. H. Holmes.—First seen April 27, 1885; next seen April 29, common April 30; breeds abundantly.

Walla Walla, W. T. J. W. Williams.—May 20 four birds, first seen; next seen June 2; June 10 common; tolerably common in breeding season; young seen June 30.

British Columbia. John Fannin.—Common summer resident.

Ridgway. Truckee Valley (not far from the eastern base of the Sierra). One specimen seen June 2, 1868; not again met with in Nevada. Dr. Cooper (Proceedings Nat. Mus. 1879), says it was first seen at San Diego April 25, 1862—a backward spring like that of 1884; Haywards, April 20, 1875; April 23, 1876.

287. Turdus ustulatus swainsonii (Cab.) OLIVE-BACKED THRUSH.

British Columbia. John Fannin. — An abundant summer resident.

East Humboldt Mountains. Ridgway.—Encountered in considerable numbers during the season of their southward migration.

Pine forests of Calaveras County. L. B.—Summer, breeding, common in several localities, probably common in numerous localities in the coniferous forests of the Sierra, in summer, from Tuolumne County northward. (See Proc. Cal. Acad. Ser. 2, ii, 60.)

288. Turdus aonalaschkæ Gmel. DWARF HERMIT THRUSH.

San Diego. L. B.—Common winter visitant; April 12, 1884, two birds, the last seen.

Poway. F. E. Blaisdell.—Common winter visitant; not noticed after April 1; first seen the following fall, October 24, 1884; remaining until April 8, 1885.

Volcan Mountains. W. O. Emerson.—Rare to February 22, 1884, when it apparently left the mountain in consequence of severe weather.

Mount Whitney. Henshaw, 1876.—None were seen previous to the very last of September. After this time every little willow thicket along the mountain streams contained one or more. The migration was at its height from the 5th to the 15th of October.

Oakland and vicinity. W. E. Bryant.—Common winter visitant.

Olema. A. M. Ingersoll.—Common to April 15; last seen May 3.

Stockton. L. B.—April 25, 1879, last seen; season backward.

Beaverton. A. W. Anthony.—Common summer resident; first nest seen June 7, 1884.

Walla Walla, W. T. Dr. Williams.—April 27, 1885; common May 15; young seen June 30.

British Columbia. John Fannin.—Summer resident; not common.

Henshaw, 1879. By the last of August it was numerous along the foothills of the Cascade range of Oregon.

Ridgway. But one individual was met with, this one being secured. [The specimen in question was obtained on Trout Creek, a tributary of the Humboldt River.]

Berkeley. T. S. Palmer.—October 12, 1885, first seen; next seen October 19, and already common; last seen April 4, 1886. It is an abundant winter resident; does not breed.

Haywards. Dr. Cooper (Proc. Nat. Mus., 1879). Winter visitant; arrived October 12, 1875; left May 1.

Saticoy. Dr. Cooper (Proc. Nat. Mus., 1879). Arrive November 5, 1872.

289. **Turdus sequoiensis** Belding. BIG TREE THRUSH.

Sierra Nevada Mountains, San Bernardino, Calaveras, Placer counties (Coast Range, Monterey Co.?)

I think the specimen from San Bernardino Mountains that I saw, collected by Mr. Price, identical with specimens I got at Calaveras and Placer counties, but hardly know what to think of the specimens Mr. Bryant got near Monterey in the summer of 1889. They may be like those Dr. Merrill got at Fort Klamath (See Auk, 1888, page 365), and there may be still another undescribed, small, pallid thrush, that breeds in California. More specimens and observations are much needed to determine some of the difficult questions concerning these birds.

[Upon a request from Mr. Belding I have measured all the available small thrushes which are referable to *T. sequoiensis* and present the dimensions here.—W. E. B.]

No.	Sex	Locality. California.	Date. 1889.	Wing.	Tail Feathers	Bill from Nostril.	Tarsus.	Middle Toe and Claw.	
				mm	mm	mm	mm	mm	
326	♂	Big Trees	May 26	94	68	9	28.5	22	Coll. Cal. Acad. Type.
788	♂	San Bernardino Mts	July 2	96	68	10.5	29	22	Coll. W. W. Price.
3603	♂	Monterey Co	July 3	87	64	9.5	29	19	Coll. W. E. Bryant.
3604	♂	Monterey Co	July 3	83	67	9.5	28	20.5	Coll. W. E. Bryant.
3605	♂	Monterey Co	July 5	81.5	63.5	10	29	.0	Coll. W. E. Bryant.
3606	♂	Monterey Co	July 5	87	65	9.5	28	20	Coll. W. E. Bryant.
3607	♀	Monterey Co	July 5	81	62	10	28	20.5	Coll. W. E. Bryant.
327	♀	Big Trees	May 23	91	67	10.5	28.5	22	Coll. Cal. Acad. Type.
374	♀	Lake Tahoe	August 12	88.5	66	10.5	27	20	Coll. Cal. Acad.
787	♀	San Bernardino Mts	July 2	92.5	67	10.5	28	21	Coll. W. W. Price.

290. Turdus aonalaschkæ auduboni (Baird). AUDUBON'S HERMIT THRUSH.

Julian, San Diego County. N. S. Goss.—March 17.

San Bernardino Mountains. F. Stephens.—Upper pine regions 9,000 feet altitude breeds.

Later, Mr. Stephens wrote: "I am not certain whether the eggs I got in the San Bernardino Mountain are *auduboni* or *nanus*, as I wounded the parent bird but did not get her. On looking over my skins I find three, one of which is *nanus*, one is var. *auduboni*—both were taken near Campo—the third is very nearly intermediate between *nanus* and *auduboni*."

Henshaw, 1879. During the summer of 1877 I heard in several of the subalpine valleys of northeastern California what were without doubt the Audubon's thrushes but failed to secure specimens. Here they were evidently not very numerous, but in the mountains back of Camp Bidwell the succeeding season the same thrush was heard and satisfactorily identified by shooting the bird. They were here very abundant.

Santa Barbara. J. Amory Jeffries.—(Auk, 1888, page 222.) Came April 2, 1883.

L. B.—Perhaps all of the above notes refer to the bird I described in these Proceedings, June 11, 1889, under the name *Turdus sequoiensis*.

291. Merula migratoria propinqua Ridgw. WESTERN ROBIN.

San Diego. L. B.—Common in the winter of 1883–84. It disappeared March 22 but returned on the 29th and remained two days while the mountains east of San Diego were covered with snow. One was seen at Campo, May 14, just previous to a rainstorm. No robins were seen during a four days' journey south of Campo, about half of which was through pine-clad mountains, from May 9 to 13. Mr. T. E. Wadham collected it in Lower California about ninety miles southeast of San Diego, April 22. His specimens showed no leaning toward var. *confinis* of the Cape region.

Poway. F. E. Blaisdell.—Common winter visitant, last seen March 30, 1884.

Poway. W. O. Emerson.—Seen April 27, Volcan Mountains, a few noticed every day.

San Bernardino. F. Stephens.—Irregular winter visitant to the valley, a rare summer resident in the mountains; Agua Caliente, March 25–28, a few seen, perhaps migrants. Common from March 18 to April 15, 1886.

Santa Barbara. Dr. William Finch.—Very abundant upon my arrival here January 1, 1876. Rainfall of that winter about thirty inches. Since then they have appeared in numbers corresponding with the quantity of rain each season. In the winter of 1877 but four and one-half inches of rain fell and not a robin put in an appearance. This winter. 1883–84, they came earlier, in greater numbers than ever, and our dry season (dry until January 25), is proving wet, wetter, wettest.

San Jose. A. L. Parkhurst.—Arrived in November; March 22, large flocks.

Berkeley. T. S. Palmer.—Began to sing March 5, 1886; last seen April 3.

Alameda and Contra Costa counties. W. E. Bryant.—Some winters abundant, others rare.

Olema. A. M. Ingersoll.—Last seen May 24, 1884.

Galt, Sacramento County. Miss Genevieve Harvey.—Robins were abundant here this winter.

Marysville. W. F. Peacock.—The bulk departed April 10; last seen April 22.

Chico. William Proud.—March 17 and 18 robins are leaving us; last seen June 26, a solitary male bird who made things ring in the garden by his lively notes. After staying two hours he headed for the mountain.

Igo, Shasta County. E. L. Ballou.—Singing at evening since April 17.

Beaverton, Oregon. A. W. Anthony.—A few were here when I arrived, February 2; about February 25 they began to arrive from the south and by March 12 were all here. First nest seen April 24.

Seattle. O. B. Johnson.—May 1, nest and two eggs.

British Columbia. John Fannin.—An abundant resident. The bulk winter on Vancouver Island. Probably a few go south but the numbers do not appear to be much less in winter than in summer.

Cape Beale, British Columbia. Emanuel Cox.—Always here.

Cape Flattery Light. Alexander Sampson, keeper.—Occasionally some come from the main land, three-fourths of a mile off, in January and in summer.

Admiralty Head, Whidby Island. S. L. Wass.—First seen February 1, 1885, one bird.

Yakima, W. T. Samuel Hubbard, Jr.—One seen February 15, 1885.

Walla Walla, W. T. Dr. J. W. Williams.—First seen March 20 (twenty birds), common April 1, 1885; common in breeding season; young seen May 20, 1885.

Beaverton, Oregon. A. W. Anthony.—One seen January 17, 1885; again January 21; common February 10; common in breeding season.

Cape Foulweather. S. L. Wass.—First seen February 22; next seen February 28, 1885; not absent more than two months, and some winters only during a cold wave.

Ukiah. George E. Aull.—Abundant resident.

Sebastopol. F. H. Holmes.—A winter visitant; a few breed but the bulk depart in April.

L. B.—Red Bluff, February 3 and 5, a single flock.

Gridley, October 1 and 3, snow on both sides of the valley above about 3,000 feet; not seen again here this winter (1884-85). Oroville, January and December, rare. Colfax and Alta very rare in the middle of November. Summit and Donner Lake, not seen from November 13 to 17, although it was mild and scarcely any snow had fallen. It is a common summer resident in the fir forests of California from latitude 38° northward; never breeding in orchards or about settlements in California, I believe. At Blood's, altitude 7,200 feet, young did not begin to leave their nests until July 14, 1880, nearly a month later than on the following year, owing to difference in the winter's snowfall and consequent difference in the advent of summer.

Marysville. W. F. Peacock.—First seen November 2; next seen December 22; common March 1. The abundance owing to the season.

Berkeley. T. S. Palmer.—Rare this year (1885); Abundant winter of 1883-84.

Alameda. H. R. Taylor.—Six seen February 16, 1885; common February 23.

Haywards. W. O. Emerson.—First seen November 29; next seen December 22; last seen April 8, 1885; rare this winter; singing in April.

Poway. F. E. Blaisdell.—First seen November 1, 1884; common February 6, 1885: rare this winter. Volcan Mountains November 15, 1884, first seen; November 16, three large flocks.

Henshaw, 1879. Found throughout this whole region as a summer visitant and is more or less abundant according to special locality. The species begins to lay in the neighborhood of Carson about the middle of May. Robins were fairly numerous in Oregon along the Columbia River during the last of October, and a few doubtless winter even at this high latitude.

Camp Harney. Bendire.—An abundant summer resident, breeding in great numbers in this vicinity. A few pass the mild winters here, frequenting at such times the junipers, whose berries furnish them their principal food. They are undoubtedly birds which have been reared farther north.

Ridgway. In the vicinity of Carson it was extremely abundant from the middle of March to the middle of April. In August they were quite plentiful in the valley of the Truckee below the "Big Bend."

Hoffman. Usually found in abundance along the timbered bottom-lands of the upper portion of the State. During breeding season occur in the timbered mountains, as at Bull Run where these birds were building during the latter part of May although the snow had not all disappeared in the ravines about the foothills.

292. **Hesperocichla nævia** (Gmel.) VARIED THRUSH.

British Columbia. John Fannin.—Common resident (1884).

Walla Walla, W. T. J. W. Williams.—Seen in this vicinity this summer (1885).

Beaverton, Or. A. W. Anthony.—Usually common in winter. March 16, 1885.

Ukiah. George E. Aull.—Common winter visitant.

Sebastopol. F. H. Holmes.—Abundant winter visant.

Olema. A. W. Ingersoll.—April 4, 1884, last one seen.

Berkeley. T. S. Palmer.—December 31, one seen; very rare (1884). Last seen March 27, 1885. April 3, 1886, abundant; usually a very rare winter visitor.

Sebastopol. F. H. Holmes.—March 19, 1885.

Santa Cruz. A. M. Ingersoll.—October 30, 1885, first seen.

Alameda and Contra Costa counties. W. E. Bryant. Rare winter visitant.

Haywards. W. O. Emerson.—December 19, 1884, one; rare.

San Jose. A. L. Parkhurst.—Only one or two this winter (1884).

Ventura County. Evermann.—But one seen.

San Bernardino Mountains. F. Stephens.—Rare winter visitant.

Volcan Mountains. F. E. Blaisdell.—November 15, a pair; November 25, one seen.

Alta and Colfax, November 17–18 (1884), rare.—L. B.

Alameda. H. R. Taylor.—March 3, 1885, four in foothills; March 16, one.

Summit, Central Pacific Railroad, October 5, two seen; common October 10, 1885.—L. B.

Soda Springs (ten miles south of Summit) L. B.— October 1, 1877, three adult male specimens; usually common in the foothills of Central California in winter, also in the valleys in suitable localities; rare during the milder winters, as in 1884–'85.

Berkeley. T. S. Palmer.—First seen November 7, 1885; next seen November 11; common November 13; usually rare.

Henshaw, 1879. From Mr. H. G. Parker I have information of the occurrence of this thrush near Reno in western Nevada. Very large numbers made their appearance about February 1, and remained until into March.

293. Saxicola œnanthe (Linn.) WHEATEAR.

British Columbia, John Fannin.—Very rare; breeds east of the Cascades.

294. **Sialia mexicana** Swains. WESTERN BLUEBIRD.

Seattle. O. B. Johnson.—First seen March 3, 1884.

British Columbia. John Fannin.—Abundant summer resident. March 10, 1885, first seen; March 12, next seen; common April 6; breeds.

Henshaw, 1879. East slope from Carson to the Columbia River—the common bluebird of the region.

Camp Harney. Bendire.—Common during their migrations. None are known to remain and breed.

Ridgway. We lost sight of the species entirely after we left the eastern watershed of the Sierra Nevada going eastward. Carson, February 21, 1868, two specimens.

Fort Walla Walla. J. W. Williams.—April 4, two arrived, the first of the season. It does not winter here.

San Diego. L. B.—Rather common resident breeding in the timbered parts of the country. It is more common near San Diego in winter than in summer.

Poway. F. E. Blaisdell.—Seen in flocks to February 11, 1884, afterward in pairs.

Volcan Mountains. W. O. Emerson.—Seen every day, pairing by March 1, 1884; the female then looking for nesting places; at this time common.

San Bernardino. F. Stephens.—A tolerably common visitant in winter to the valley, breeding in the mountains. Agua Caliente, common March 25. Seen almost every day from March 18 to April 15, 1886.

Berkeley. T. S. Palmer.—I know of only one instance of its breeding here. Last seen March 18, 1886.

Alameda and Contra Costa counties. W. E. Bryant.—Tolerably common resident.

Ukiah. George E. Aull.—Common resident.

Central California. L. B.—Common resident. Not breeding much above 5,000 feet; tolerably common at Red Bluff, February 3, 1885.

Beaverton, Oregon. A. W. Anthony.—Common summer resident; first seen February 29, 1884. March 15 abundant.

295. Sialia arctica (Swains.) MOUNTAIN BLUEBIRD.

San Diego. L. B.—Common until March 15, 1884, when it disappeared, but a large flock returned March 29 during a cold rain-storm and stayed two days. April 4 I shot an apparently healthy female, the last seen. It breeds commonly about mountain meadows in Calaveras, Alpine, Placer and Butte counties, and no doubt has a much more extended breeding range on the Sierra, both north and south. In Calaveras County it does not appear to breed below 7,000 feet, above the breeding range of *S. mexicana*, though I found a few pairs of both species breeding at an altitude of 5,800 feet in Butte County, an unusual occurrence.

Poway. F. E. Blaisdell.—It arrived November 10, 1884, was last seen February 11. This is the first I have seen of the species here, during a residence of ten years. I think I should have noticed them before if they had been in the habit of visiting this locality. January 3, 1885, a flock seen; the only ones seen this winter.

Alameda and Contra Costa counties. W. E. Bryant. Tolerably common winter visitant.

Marysville. W. F. Peacock.—February 13, 1884, last seen.

British Columbia. John Fannin.—Rare migrants; found only east of Cascade Mountains.

Ridgway. This is the characteristic bluebird of the interior, and it is most numerous where other species are rarest. In June it was common at Virginia City where it nested in the manner of the eastern species in suitable places about buildings in the town. It was also common under similar circumstances at Austin, while on the higher portions of the West Humboldt, Ruby and East Humboldt mountains it was still more abundnat.

Camp Harney, Bendire.—Breeds here but is not common. In the vicinity of Cañon City, Oregon, I found it rather abundant.

Hoffman. Distributed over the northern and middle portions of Nevada. None were observed south of Hot Springs Cañon, sixty miles northwest of Belmont.

APPENDIX.

Observations on the fall migrations of birds are so seldom made that it is thought best to publish the following notes which were not inserted in their proper place.

Mr. Fannin wrote from Burrard Inlet, September 29, 1885: "Nearly all our summer residents are gone. A few may remain about Frazer River, but this place is about deserted."

Dr. Williams, at Walla Walla, said that *Dendroica æstiva, Geothlypis trichas occidentalis, Icteria virens longicauda, Chelidon erythrogaster, Habia melanocephala, Passerina amœna, Tyrannus tyrannus, Tyrannus verticalis*, and other summer residents were absent from that locality as early as August 25, 1884.

Anthus pensilvanicus arrived there two days later.

I made continuous observations at the summit of the Central Pacific Railroad from August 1, 1885, to October 12, of the same year. This is an excellent station for observing migrants. The least height of the summit is about 7,000 feet, the adjacent peaks being about two thousand feet higher. About the middle of August a general southward or westerly movement of the fall migrants began, unaccompanied with any marked change of temperature. Many Clarke's crows and Lewis's woodpeckers began to migrate as early as August 16, the former following the divide, the latter taking their usual southwest course, which would lead them to the California valleys where they were undoubtedly going, neither species being turned from its course by the highest peaks.

This southward migration of Clarke's crow continued into October and was a surprise to me, as I had previously supposed that this bird was a non-migrant.

The sparrows, warblers and hummingbirds followed the course of the divide, or nearly so, but a number of species went directly toward the Sacramento Valley. Among the latter were two species of blackbirds, a few ravens and white geese. California jays, kingfishers and several species of small birds, or rather individuals of these, were noted and were thought to be on their way down the west slope.

The first rain of the season fell on September fourth. The third and the fourth were cooler and cloudy when Say's chipmunk and the yellow-bellied marmot were evidently preparing for the storm by carrying bedding into their burrows. No marmots were seen after the rain of the fourth, and hummingbirds were rarely seen afterward. No tyrant flycatchers were seen after the ninth.

On the tenth *Anthus pensilvanicus* and *Otocoris alpestris strigata* arrived from the north.

Zonotrichia leucophrys intermedia came two days later, but *Z. coronata* did not arrive until the twenty-fifth, when there was rain and snow.

Helminthophila celata lutescens, *H. ruficapilla gutturalis*, *Geothlypis macgillivrayi*, *Spizella breweri* and *Pipilo chlorurus* were seen on September twenty-fourth, but not afterward. *Zonotrichia leucophrys* was last seen on the twenty-eighth, when *Passerella iliaca unalaschcensis* was first seen. By October tenth the migration was about over, a migration characterized mostly by short flights; made leisurely, often by single individuals acting independently. No night migrations were detected.

At Murphys, altitude about 2,300 feet, Mr. J. J. Snyder noted the female Louisiana tanager as very common; the male very rare on August 24, 1885, and that Bullock's oriole and the black-headed grosbeak were also very rare. Cliff swallows were not seen after August 18.

Violet-green swallows were numerous at three localities between Murphys and Big Trees, September 26.

At Berkeley, *Petrochelidon lunifrons* was last seen by Mr. Palmer on August 5, 1885. *Piranga ludoviciana*, September 11; *Habia melanocephala*, August 26; *Tyrannus verticalis*, three days later; *Contopus richardsonii* September 12. *Turdus ustulatus* one day later.

I was at Weber Lake in Henness Pass, altitude about 7,000 feet, from July 30, 1889, to August 7, and noticed the following, which I thought were already on their way to the Pacific Coast. The pelicans and gulls were seen to cross the mountain, going in a southwesterly direction. July 31, a large flock of juvenile *Larus californicus* was seen. August 3, a large flock of juvenile *Recurvirostra americana;* a flock of juvenile *Phalaropus lobatus;* a flock of old and young or juvenile *Spatula clypeata*, and from fifty to one hundred *Pelecanus erythrorhynchus* in a single flock. On the fifth, a flock of about a dozen *Himantopus mexicanus*, and about the same number of *Sterna forsteri* appeared. The young of *Gallinago delicata*, which bred there, besides the young of numerous species, were capable of migrating by the first of August.

According to Ridgway's Manual *Phasianus torquatus*, *P. versicolor* and *P. sœmmerringii* have been introduced into Oregon from China and Japan and are said to be doing well, and Mr. Anthony informs me that a considerable number of European species have quite recently been introduced about Portland.

Since the first part of 1886, when I considered this catalogue of the land birds of the district quite complete, several accidental occurrences have been reported, some of which were probably wrongly identified. Mr. Evermann's list of the birds of Ventura County contains several errors of identification, some of which

Dr. Cooper mentioned in The Auk, iv, 85, as errors or probable errors, to which I will add *Ammodramus beldingi*, *Dendroica graciœ*, *Parus atricapillus occidentalis*, and *Melospiza fasciata guttata*, erroneously said to be residents.

L. BELDING.

August 7, 1890.

ERRATA.

Page 1. Fourth paragraph should be inserted at foot of page 7.
Page 12. For "F. C. Blaisdell" read "F. E. Blaisdell."
Page 36. For "Techachapi" read "Tehachapi."
Page 50. For "Mulato" read "Nulato."
Page 216. For "Seimrus" read "Seiurus."

INDEX.

Acanthis linaria	135
Accipiter atricapillus striatulus	32
cooperi mexicanus	31
velox rufilatus	30
Agelaius gubernator	121
phœniceus	120
tricolor	122
Ammodramus bairdii	146
rostratus	145
sandwichensis	142
alaudinus	143
Bryanti	144
savannarum perpallidus	146
Ampelis cedrorum	195
garrulus	195
Amphispiza belli	161
nevadensis	162
bilineata	161
Anthus pensilvanicus	222
Aphelocoma californica	110
insularis	111
woodhousei	110
Aquila chrysaëtos	39
Archibuteo ferrugineus	38
lagopus sancti-johannis	37
Asio accipitrinus	48
wilsonianus	47
Auriparus flaviceps	245
Blackbird, bicolored	121
Brewer's	128
red-winged	120
tricolored	122
yellow-headed	118
Bluebird, mountain	263
western	262
Bobolink, western	117
Bob-white	8
Bonasa umbellus sabini	17
umbelloides	17
Bubo virginianus saturatus	53
subarcticus	52
Bunting, lark	180
lazuli	178
Bush-tit	243
Californian	244
lead-colored	245
Buteo abbreviatus	36
borealis calurus	33
lineatus elegans	34
swainsoni	36
Calamospiza melanocorys	180
Calcarius lapponicus	140
Callipepla californica	12
vallicola	12
gambeli	14
Campylorhynchus brunneicapillus	228
Caracara, Audubon's	46
Cardinal	175
Cardinalis cardinalis	175
Carpodacus cassini	132
mexicanus frontalis	133
purpureus californicus	131
Catbird	226
Cathartes aura	26
Catherpes mexicanus conspersus	230
Centrocercus urophasianus	19
Ceophlœus pileatus	68
Certhia familiaris occidentalis	235
Ceryle alcyon	58
cabanisi	59
Chætura vauxii	79
Chamæa fasciata	242
henshawi	243
Chat, long-tailed	219
Chelidon erythrogaster	186
Chickadee Californian	242
chestnut, backed	242
mountain	241
Oregon	240
Chondestes grammacus strigatus	147
Chordeiles texensis	78
virginianus henryi	77
Cinclus mexicanus	223
Circus hudsonius	28
Cistothorus palustris paludicola	234
Clivicola riparia	194
Coccothraustes vespertina	130
Coccyzus americanus occidentalis	57
Colaptes auratus	72
cafer	73
saturatior	75
chrysoides	75

Colinus virginianus	8	Falcon, Peale's	43
Columba fasciata	20	prairie	41
Columbigallina passerina	24	Finch, California purple	131
Contopus borealis	96	Cassin's purple	132
richardsonii	98	house	133
Corvus americanus	113	Flicker	72
cauriuus	115	gilded	75
corax sinuatus	112	northwestern	75
Cowbird	117	red-shafted	73
dwarf	118	Flycatcher, acadian	100
Creeper, California	235	ash-throated	93
Crossbill, American	134	Hammond's	102
Crow, American	113	little	101
northwest	115	olive-sided	96
Cuckoo, California	57	vermilion	105
Cyanocephalus cyanocephalus	116	western	99
Cyanocitta stelleri	109	Wright's	103
frontalis	109	Galeoscoptes carolinensis	226
Cypseloides niger	79	Geococcyx californianus	56
Dendragapus franklinii	17	Geothlypis macgillivrayi	216
obscurus fuliginosus	15	trichas occidentalis	218
richardsonii	16	Glaucidium gnoma californicum	55
Dendroica æstiva	208	Gnatcatcher, black-tailed	249
auduboni	210	plumbeous	248
cærulescens	209	western	248
coronata	209	Goldfinch, American	136
maculosa	212	Arkansas	137
nigrescens	212	Lawrence's	138
occidentalis	215	Goshawk, western	32
townsendi	213	Grosbeak, black-headed	175
Dipper, American	223	evening	130
Dolichonyx oryzivorus albinucha	117	pine	131
Dove, ground	24	western blue	177
mourning	22	Grouse, Columbian sharp-tailed	18
white-winged	23	Franklin's	17
Dryobates nuttallii	62	gray ruffed	17
pubescens	60	Oregon ruffed	17
gairdnerii	60	Richardson's	16
scalaris bairdi	61	sage	19
villosus harrisii	59	sooty	15
Eagle, bald	40	Habia melanocephala	175
golden	39	Haliæetus leucocephalus	40
Ectopistes migratorius	21	Harporhynchus crissalis	228
Elanus leucurus	27	lecontei	227
Empidonax acadicus	100	redivivus	227
difficilis	99	Hawk, American rough-legged	37
hammondi	102	American sparrow	44
pusillus	101	duck	42
wrightii	103	marsh	28
Falco columbarius	43	pigeon	43
suckleyi	44	red-bellied	34
mexicanus	41	Swainson's	36
peregrinus anatum	42	western Cooper's	31
pealei	43	western red-tailed	33
richardsonii	44	western sharp-shinned	30
sparverius	44	zone-tailed	36

INDEX. 271

Helminthophila celata 206
            lutescens 206
            luciæ 205
            ruficapilla gutturalis 205
            virginiæ 205
Hesperocichla nævia 260
Hummingbird, Allen's 88
            Anna's 84
            black-chinned 82
            broad-tailed 85
            calliope 80
            Costa's 83
            Floresi's 85
            rufous 85
            violet-throated 83
Icteria virens longicauda 219
Icterus bullocki 126
            cucullatus nelsoni 125
            parisorum 125
Jay, blue-fronted 109
            California 110
            Oregon 111
            piñon 116
            Santa Cruz Island 111
            Steller's 109
            Woodhouse's 110
Junco hyemalis 159
            oregonus 159
            Oregon 159
            slate-colored 159
Kingbird 89
            Arkansas 90
            Cassin's 92
Kingfisher, belted 58
            Texan 59
Kinglet, ruby-crowned 246
            western golden-crowned 245
Kite, white-tailed 27
Lagopus leucurus 18
            rupestris 18
Lanius borealis 197
            ludovicianus gambeli 198
Lark, Mexican horned 105
            pallid horned 105
            ruddy horned 105
            streaked horned 106
Leucosticte, gray-crowned 135
            Hepburn's 135
            tephrocotis 135
            littoralis 135
Longspur, Lapland 140
Loxia curvirostra minor 134
Magpie, American 107
            yellow-billed 108
Martin, western 183
Meadowlark, western 123

Megascops asio bendirei 51
            kennicottii 51
            flammeolus 52
Melanerpes formicivorus bairdi 69
            torquatus 70
            uropygialis 72
Melopelia leucoptera 23
Melospiza fasciata guttata 166
            heermanni 164
            montana 163
            rufina 167
            samuelis 165
            lincolni 167
            striata 168
Merlin, black 44
            Richardson's 44
Merula migratoria propinqua 256
Micropallas whitneyi 56
Mimus polyglottos 226
Mniotilta varia 205
Mockingbird 226
Molothrus ater 117
            obscurus 118
Myadestes townsendii 250
Myiarchus cinerascens 93
Nighthawk, Texan 78
            western 77
Nutcracker, Clarke's 115
Nuthatch, pigmy 238
            red-breasted 237
            slender-billed 236
Nyctala acadica 50
Nyctea nyctea 53
Oreortyx pictus 9
            plumiferus 9
Oriole, Arizona hooded 125
            Bullock's 126
            Scott's 125
Oroscoptes montanus 224
Osprey, American 46
Otocoris alpestris chrysolæma 105
            leucolæma 105
            rubea 105
            strigata 106
Owl, American barn 47
            American hawk 54
            American long-eared 47
            burrowing 54
            California pigmy 55
            California screech 51
            dusky horned 53
            elf 56
            flammulated screech 52
            great gray 60
            Kennicott's screech 51
            saw whet 50

Owl, short-eared	48	Poor-will, California	75
snowy	53	Progne subis hesperia	183
spotted	49	Psaltriparus minimus	243
western horned	52	californicus	244
Pandion haliaëtus carolinensis	46	plumbeus	245
Partridge, California	12	Pseudogryphus californianus	24
Gambel's	14	Ptarmigan, rock	18
mountain	9	white-tailed	18
plumed	9	Pyrocephalus rubineus mexicanus	105
valley	12	Raven, American	112
Parus atricapillus occidentalis	240	Redpoll	135
gambeli	241	Redstart, American	222
inornatus	240	Regulus calendula	246
griseus	2 0	satrapa olivaceus	245
rufescens	242	Road-runner	56
neglectus	242	Robin, western	256
Passer domesticus	106	Rough-leg, ferrugineous	38
Passerella iliaca	169	Salpinctes obsoletus	229
megarhyncha	170	Sapsucker, red-breasted	66
schistacea	171	red-naped	65
unalaschcensis	169	Williamson's	67
Passerina amœna	178	Saxicola œnanthe	261
Pediocætes phasianellus columbianus	18	Sayornis nigricans	95
Perisoreus obscurus	111	saya	94
Petrochelidon lunifrons	184	Scolecophagus cyanocephalus	128
Peucæa ruficeps	163	Scotiaptex cinerea	50
Phalænoptilus nuttalli californicus	75	Seiurus noveboracensis notabilis	216
Phasianus sœmmerringti	168	Setophaga ruticilla	222
torquatus	168	Shrike, California	198
versicolor	168	northern	198
Phœbe, black	95	Sialia arctica	292
Say's	94	mexicana	293
Pica nuttalli	108	Siskin, pine	139
pica hudsonica	107	Sitta canadensis	237
Picicorvus columbianus	115	carolinensis aculeata	236
Picoides americanus dorsalis	65	pygmæa	238
arcticus	64	Snowflake	140
Pigeon, band tailed	20	Solitaire, Townsend's	250
passenger	21	Sparrow, Baird's	146
Pinicola enucleator	131	Belding's marsh	144
Pipilo aberti	175	Bell's	161
chlorurus	173	black-chinned	158
fuscus crissalis	174	black-throated	161
maculatus megalonyx	171	Brewer's	157
oregonus	172	Bryant's marsh	144
Pipit, American	222	English	168
Piranga ludoviciana	181	Forbush's	168
rubra cooperi	182	fox	169
rubriceps	182	Gambel's	151
Plectrophenax nivalis	140	golden-crowned	153
Polioptila cærulea obscura	248	Heermann's song	164
californica	249	intermediate	150
plumbea	248	large-billed	145
Polyborus cheriway	46	Lincoln's	167
Poocætes gramineus affinis	142	mountain song	163
confinis	140	Oregon vesper	142

INDEX. 273

Sparrow, rufous crowned 163
   rusty song 166
   sage 162
   Samuels's song 165
   Sandwich 142
   slate-colored 171
   sooty song 167
   thick-billed 170
   Townsend's 169
   western chipping 155
   western grasshopper 146
   western lark 147
   western savanna 143
   western tree 155
   western vesper 140
   white-crowned 148
   white-throated 154
Speotyto cunicularia hypogæa 54
Sphyrapicus ruber 66
   thyroideus 67
   varius nuchalis 65
Spinus lawrencei 138
   pinus 139
   psaltria 137
   tristis 136
Spizella atrigularis 158
   breweri 157
   monticola ochracea 155
   socialis arizonæ 155
Stelgidopteryx serripennis 193
Strix pratincola 47
Sturnella magna neglecta 123
Surnia ulula caparoch 54
Swallow, bank 194
   barn 186
   cliff 184
   rough-winged 193
   tree 189
   violet-green 191
Swift, black 70
   Vaux's 79
   white-throated 80
Sylvania pusilla 220
   pileolata 220
Syrnium occidentale 49
Tachycineta bicolor 189
   thalassina 191
Tanager, Cooper's 182
   Gray's 182
   Louisiana 181
Thrasher, Californian 227
   crissal 228
   Leconte's 227
   sage 224
Thrush, Audubon's hermit 256
   big tree 255

Thrush, dwarf hermit 254
   olive-backed 253
   russet-backed 251
   varied 260
Thryothorus bewickii spilurus 231
Towhee, Abert's 175
   Californian 174
   green-tailed 173
   Oregon 172
   spurred 171
Trochilus alexandri 82
   alleni 88
   anna 84
   calliope 89
   costæ 83
   floresii 85
   platycercus 85
   rufus 85
   violajugulum 83
Troglodytes aëdon parkmanii 232
   hiemalis pacificus 234
Turdus aonalaschkæ 254
   audoboni 256
   sequoiensis 255
   ustulatus 251
   swainsonii 253
Tyrannus tyrannus 89
   verticalis 90
   vociferans 92
Verdin 245
Vireo belli pusillus 204
   blue-headed 201
   Cassin's 201
   flavoviridis 199
   gilvus 199
   gray 204
   huttoni 203
   Hutton's 203
   least 204
   olivaceus 199
   plumbeous 202
   red-eyed 199
   solitarius 201
      cassinii 201
      plumbeus 202
   vicinior 204
   warbling 199
   yellow-green 199
Vulture, California 24
   turkey 26
Warbler, Audubon's 210
   black and white 205
   black throated blue 209
   black-throated gray 212
   Calaveras 205
   hermit 215

Warbler, Lucy's.................... 205
    lutescent........................ 206
    macgillivray's................... 216
    magnolia......................... 212
    myrtle........................... 209
    orange crowned................... 206
    pileolated....................... 220
    Townsend's....................... 213
    Virginia's....................... 205
    Wilson's......................... 220
    yellow........................... 208
Water thrush, Grinnell's............ 216
Waxwing, Bohemian................... 195
    cedar............................ 195
Wheatear............................ 261
Woodpecker, alpine, three-toed...... 65
    arctic three-toed................ 64
    Baird's.......................... 61
    Californian...................... 69
    downy............................ 60
    Gairdner's....................... 60
    Gila............................. 72
    Harris'.......................... 59

Woodpecker, Lewis's................. 70
    nuttall's........................ 62
    pileated......................... 68
    white-headed..................... 63
Wren, cactus........................ 229
    cañon............................ 230
    Parkman's........................ 232
    rock............................. 229
    tule............................. 234
    Vigor's.......................... 231
    western winter................... 234
Wren-tit............................ 242
    pallid........................... 243
Xanthocephalus xanthocephalus....... 118
Xenopicus albolarvatus.............. 63
Yellow-throat, western.............. 218
Zenaidura macroura.................. 22
Zonotrichia albicollis.............. 154
    coronata......................... 153
    leucophrys....................... 148
        gambeli...................... 151
        intermedia................... 150

www.ingramcontent.com/pod-product-compliance
Lightning Source LLC
Chambersburg PA
CBHW031933230426
43672CB00010B/1907